Merry ❄ Xmas

from
John & Marian

CUTTING UP THE NORTH

CUTTING UP THE NORTH

*The History of the Forest Industry
in the Northern Interior*

By: Ken Bernsohn

ISBN 0-88839-114-5

Canadian Cataloguing in Publication Data

Bernsohn, Ken, 1943-
   Cutting up the North

   Bibliography: p.
   Includes index.
   ISBN 0-88839-114-5

   1. Lumber trade - British Columbia - History. 2. Logging - British Columbia - History. 3. Forests and forestry - British Columbia - History. I. Title.
HD9764.C33B753   338.4'7674'0097111   C81-091157-4

All rights reserved. No part of this publication may be reproduced, stored in a retrieval system or transmitted, in any form or by any means, electronic, mechanical, photocopying, recording or otherwise, without the prior written permission of Hancock House Publishers.

Edited by Margaret Campbell and Nancy Flight
Design by Peter Burakoff
Layout by Peter Burakoff, Linda Rourke and Anne Whatcott
Typeset by Sandra Sawchuck, Anne Whatcott, Linda Rourke and Diana Lytwyn in Times Roman on an AM Varityper Comp/Edit.
Printed in Canada by Friesen Printers

Cover photos courtesy of B.C. Provincial Archives and B.C. Ministry of Forests.

Hancock House Publishers Ltd.
#10 Orwell Street, North Vancouver, B.C., V7J 3K1

Hancock House Publishers Inc.
1431 Harrison Ave., Blaine, Washington 98230 USA

This book is for
Kathy Bernsohn,
who made me do it.

Acknowledgments

Despite the author's name on the spine of this book, a lot of people worked very hard to develop what you're about to read.

The Canada Council provided a grant that subsidized much of the research and their help is gratefully acknowledged.

About three hundred people in the industry, the IWA, the Forest Service, government, and at universities took time to talk about their experiences in the industry. Many provided articles, booklets, and records.

In some cases the people interviewed had to put up with dumb questions, and repeated visits. Van Scoffield of Northwood patiently explained how the industry worked and why it did what it did. Undoubtedly there are errors, but Van took an incredible amount of time to try to set the record straight. Toby Mogensen of the IWA got the project started by commissioning a history of the union in the north, and by explaining for many hours a lot of things not covered in textbooks, but common in the real world. Permission to quote from *Slabs, Scabs and Skidders* is appreciated. Cameron Young, editor of *ForesTalk,* provided permission to use material originally developed for his magazine. Ray Williston, Tom Waterland, Bill Young, Doug Little, and John Whitmer all had their patience tried, and tried again. Mike Siddall, publisher of *Hiballer Forest Magazine* and *Forest Insight*, encouraged the writing of the book—without even knowing how the industry would be treated—and tolerated having a writer on staff often concerned more with 1937 than with the present.

Margaret Campbell of Hancock House now knows more about the forest industry than she ever wanted to learn, thanks to analyzing the first draft of this book, and introducing some order and logic to the presentation. Then she had the "opportunity"—as in "you have the opportunity to scrub the floor with a toothbrush"—to go over it again. And again. Her tact has been amazing. Nancy Flight of Hancock House attempted, sometimes with success, to introduce the author to grammar and consistent style. Their patience was only exceeded by that of my wife, Kathy, who lived with this project, rather than with a husband, for years.

The staffs at the Provincial Archives, UBC library, Vancouver Public Library, and especially the Prince George Library were unfailingly helpful. John Bovey, Provincial Archivist, provided permission to use excerpts from the BCR Royal Commission transcripts and final report, submissions to the Pearse Royal Commission and the oral history transcript of an interview with Ray Williston. A special word for work above and beyond the call of duty should go to Mrs. Thelma Werschke, who copied photos in the Prince George

Public Library's local history collection on short notice and on her own time.

There are some harsh things said about the forest industry and the Forest Service on the following pages. But in the north, the people involved with our forests are surprisingly frank, and more honest than one would expect. People in the north say what they believe, and you can take it or leave it. When you consider that the people interviewed had no idea how the material they provided would be used, their forthright attitude is amazing. While I've disagreed with what was done, nothing in these pages is meant either as an attack on an individual's sincerity, or as an implication that the people in the north weren't doing what they genuinely believed to be the right thing at the time.

Cutting Up the North was written with mixed emotions, with premises and beliefs changing as data was uncovered that challenged prejudices and as people explained why they did various things. All the major themes grew out of the research, rather than acting as a framework at the beginning. The people talked to, whether in the industry or, as in the case of Eli Sopow and Peter Pearse, analysts of the industry, shaped this book.

As a result it's the story of people trying to make money, for a firm or for the provincial government, all too often ignoring the effect on the forest. It tells why the goose that lays the golden eggs—the forest that provides the majority of the money in and from the north—is suffering from anemia and a variety of other illnesses.

There is a great deal of talk about "free enterprisers" in the forest industry.... Unfortunately, the term is used at times where "freebooter" would be more appropriate.... The free enterprise system is justifiable only as long as it serves the public interest.

> Honourable Robert E. Sommers
> Minister of Lands & Forest
> February 1956

Table of Contents

Acknowledgments 8
1. The First Boom 10
2. Interlude: From Speculation to Conjecture
   —The First Royal Commission 16
3. Boom to Bust: 1911 - 1914 20
4. The Years of Hope: 1914 - 1939 24
5. Frustration: The Story of The
   Forest Service 36
6. How Big is the Forest? 40
7. A Hell of a Way to Earn a Living 44
8. World War II: Father of the Union 50
9. The Second Royal Commission:
   Political Flim-Flam Strikes Again 54
10. The Postwar Boom 58
11. The Pacific and Great Eastern arrives
    —40 Years Behind Schedule 62
12. The Messiest Strike in Western Canada 66
13. Recovery and Prelude: 1954 - 1956 76
14. Farce: The Third Royal Commission 80
15. Genesis 82
16. Exodus: Competition and Small Firms
    Leave the Forest Industry 86
17. The Era of Consolidation 92
18. The Smell of Money 96
19. Big Money 104
20. The End of the Boom 110
21. Portraits of the Survivors 114
22. Another View of the Forest:
    The Eco-Freaks Meet the Dinosaurs 132
23. The Saga of the NDP 140
24. Catechism on a Jig-Saw Puzzle:
    The Fourth Royal Commission 154
25. Intermission: Meet the Cast 164
26. The High Rollers 168
27. The New Forest Act: 1978 172
28. Secrets 180
Bibliography 186
Index 190

Chapter 1
The First Boom

Fallers at work, United Grain Growers Sawmill near Dewey, 1923. Photo courtesy of B.C. Forest Service.

Today the history of the lumber industry in the north is almost invisible. All that's left are rotting piles of sawdust that mark where over a thousand sawmills used to be, the remains of plank roads, and decaying piles of timber along the Goat River left from the 1950s.

But the early days of the forest industry, with all the false starts, booms, and promised booms, established the rules of the game. Today, though giant machines snip trees like celery stalks at the rate of one every minute, the forest industry and the government still live by those rules.

Unfortunately, some of the rules aren't based on what really happened; they're based on myths. But in the north of British Columbia—where the forest industry provides about 75 cents out of every dollar—the myths are as important as the reality.

In 1805 Simon Fraser founded Fort McLeod, the first settlement in British Columbia. By 1807 Fort St. James, Fort Fraser, and Fort George were in business. Ninety-six years later they were still forts, surrounded by a straggle of homes. In the economy of the province, the north was like beer and pop bottles in a basement: something you might have to cash in someday, but of little importance. When the builders of the Canadian Pacific Railway complained, in the 1880s, that their land grant of twenty miles on each side of the line included "land useless for agriculture and other productive enterprise," the provincial government turned over 5,470 square miles to the federal government to help make up the difference. British Columbia didn't get the Peace River Block back until 1930.

During this period, when the north was where gold, furs, and some fish came from, Father A.G. Morice traveled the north for the Catholic church, drawing the first map of northern British Columbia, writing down what had happened, and working with the few settlers and the many Indian bands. After returning to Toronto, he wrote the *History of the Northern Interior of British Columbia*, published in 1904. Here's his opinion of the northern forests: "Apart from the animals to which they give shelter, these woods afford but very meagre resources adapted to the wants of man."

To the 190,000 people in British Columbia just after the turn of the century, the north was a fringe benefit. The beer bottles in the basement could keep gathering dust.

But by the time Father Morice's book came out, it was out of date. Sir Richard McBride had been elected the premier of British Columbia in 1903, and he changed the situation, starting the first northern boom, in a single year.

The government was in debt. Raising taxes would be unpopular. McBride felt he had to find a way to raise money without raising taxes, and he found it by modifying what the previous government had done. Instead of selling land outright to timber companies, which left the timber firms with logged-over land they didn't want, the government had sold something it called "Special Licences" that let firms cut what they wanted from a thousand acres for a year. These had been a lot more popular than previous government attempts to make money from the forest, such as Timber Leases, Pulp Leases, and Hand Loggers' Licences. But from a political point of view, the Special Licences had some problems. In order to make money from them, people had to cut the timber they'd paid for, so they only appealed to serious loggers.

McBride made Special Licences appeal to everyone and made trees pay without being cut. First he made the licences transferable and good for five years instead of one. Then, in 1905, Special Licences were changed so they could be renewed for twenty-one years. Anyone who figured there might be a need for timber in the future could tie up land—and pay the government. Lumber firms in the eastern United States and in the Midwest were running out of timber. So a wave of speculation filled the government coffers.

In 1902 nine percent of the government budget—$445,000—came from the forest. In 1908 it brought in almost $2.5 million—over forty percent of an increased budget. And McBride's policy of having the forest help the government as much as possible is still British Columbia's forest policy.

McBride and his Minister of Lands, Fred J. Fulton, had an advantage over earlier governments. American lumbermen had cleared the woods in Maine and logged the best forests in Michigan. Some Michigan firms had moved their harvesting operations to Canada and floated the logs south to keep their mills going. Then in 1897 Ontario passed a law saying any wood harvested in that province had to be milled there. Quebec and the Maritimes followed Ontario's lead. Soon operations were closed down and moved west to Oregon, Washington, and California. In 1905 the U.S. government squeezed the timber supply by withdrawing 170 million acres for national forests. The new U.S. Forest Service designated what could be cut, when, where, and by whom. This seemed an intolerable restriction to firms that had made their fortunes "opening the wilderness for farmland." So they moved north, since British Columbia didn't impose these restrictions.

To direct the American firms to the areas where the government wanted them to go, McBride and Fulton had the Forest Act changed in 1906 so any wood cut west of the Cascades had to be processed in British Columbia. This "helped" investors look beyond the Lower Mainland and Vancouver Island. The rush moved east, then north.

In 1907, the peak year of the rush, 2,124 new licenceholders appeared in the *B.C. Gazette* in the first nine months of the year. Only 77 were B.C. mill-owners. The rest were loggers, timber cruisers, professional speculators, real estate agents, company agents, miners, bookkeepers, plus 258 Americans.

Mountain tops, swamps, areas without any trees, and good forests were included in overlapping claims. In 1910 the first Royal Commission on Forests estimated that even when all these mistakes, unsurveyed areas, and poor lands were eliminated, twice as much land was put under licence between 1905 and 1907 as the government had tied up in all other forms of tenure combined, excluding only the Canadian Pacific's railway belt.

A lot of that land was in the north. The June 1907 issue of *Western Lumberman* reported:

> ...*In the interior of the province timber cruisers are working further and further to the north all the time. Following the announcement of the construction of the Grand Trunk*

Pacific railway across British Columbia from the Rockies to the Coast, there came hundreds of applications for timber situated in districts to be traversed by the railway. In the valleys of the Upper Fraser River, in the valleys of the Endaco, the Nechaco, the Bulkley and the Skeena rivers have been posted the notices of the timber cruisers, and they also dot the shores of all other rivers and lakes, large and small, which will be given transportation facilities when the new Canadian transcontinental railway is finished.

In 1907 the Grand Trunk Pacific was in the middle of Alberta. The route through British Columbia had been roughly laid out. It was going to follow the Upper Fraser to Fort George, then cross the river and continue west just south of the Nechako. Wood was needed: timbers for tunnels, poles for telegraph lines, ties for the road bed. Men hiked in over the Yellowhead Pass from Alberta to camp ahead of the railroad and cut what was needed. Others came north by scow to Fort George, then started walking.

Both sides of the Fraser River from Rearguard Falls near the Yellowhead Pass to Fort George were taken up, then the tributaries began to be staked. Fred T. Cromwell of Portland, Oregon, staked fifty square miles on the McGregor River in an area so remote that logging didn't begin on most of it until the 1960s. On his application for a licence Cromwell listed his occupation as "capitalist."

If a man had a little money, he could stake what seemed like a sure thing. If he had some more, he could hire a timber cruiser to stake for him. If he didn't have any money at all, he could come north, ignore the rule calling for a licence, cut ties, make money, and get out.

There were more people in the woods after a quick buck than on the opening day of deer season.

Then McBride encouraged more people to come north. In 1908 the railroad began construction from the west, at Prince Rupert, as well. More people came north to cut ties, to survey the line, to clear the roadway. Bootleggers and prostitutes followed the work gangs. "Sandbar Lill," for instance, used a scow on the Fraser to bring her girls to the workers. At each camp she'd dock on the nearest sandbar and go into business. At the same time as the prostitutes came the land speculators, ready to cash in on a different, but equally basic, emotion.

In the fall of 1908 George J. Hammond staked 100 acres west of the Hudson's Bay Company trading post as the townsite for Fort George—often called Central Fort George by competing speculators. Then he incorporated the Natural Resources Security Company to sell the lots. The following spring, sixty acres next to the fort, south of Hammond's site, was bought by the Northern Development Company, backed by A.C. Hamilton with Nick Clarke as the front man. They called their new town South Fort George.

The area is a flat flood plain bordered on the north by the Nechako River, on the east and south by the Fraser River, and on the west by hills. If you face north, the original fort was about twelve o'clock high, Central Fort George was at ten, and South Fort George was at five. The crescent covered by the hands of your watch as they go from twelve to three was an Indian reserve.

If you know Prince George today, the fort was on First Street, near the CN station; Central Fort George was where the Spruceland Shopping Centre is now; and South Fort George was south of Hudson's Bay Slough, along Queensway.

Everyone knew the railroad would cross the Fraser onto the flood plain. The question was *where* it would cross. The Grand Trunk Pacific had been developing townsites as it moved west, and the promoters knew the railway

Nick Clarke's mill, Fort George, 1910. Photo courtesy of B.C. Provincial Archives.

would want to develop one in the area. They were sure it would have to be close to the already established towns, however, so one would be sure to profit. Freight would come up the Fraser and be loaded on the new railroad. There was good farming country nearby. The area was sure to boom. Clarke realized scows wouldn't be good enough, so he had a steamer built, the *Nechacco,* later renamed the *Chilco,* that ran between Soda Creek and South Fort George, carrying passengers, freight, and the machinery for a sawmill he started, the first permanent building in South Fort George. The logs for the sawmill at first came from road allowances, creating the first streets in the town in November 1909. Later the logs came from building lots, were processed through the mill, and then often went back to the same lots as boards to be used in building.

When John McInnis, MLA from the area in the late 1940s, arrived in South Fort George in March 1910, there were about thirteen permanent settlers. John lived in a tent, like the others who arrived that spring, bought boards at Clarke's mill and began to build a home so his wife could join him. More people arrived. Then a lot more. By midsummer the town had 300 residents. William Cooke, a partner in the Northern Development Company with Clarke and Hamilton, started a second sawmill and hired William Peden as manager. Unlike Clarke's river-front mill, this one was at the base of the hills to the west. Since it didn't help Central Fort George's image if settlers had to go to South Fort to get materials, the Natural Resources Security Company bankrolled a third mill, called Bogue and Brown's, on the Nechako.

By August the towns had the Bank of British North America doing business in a tent, a telephone company, a hotel, subdivisions covering 1,370 acres, and two newspapers: the *Tribune,* which had been started in South Fort but was moved to Central Fort when it was bought by Hammond, and the *Herald,* which began publication in South Fort that month.

The newspapers were supported by a provincial law requiring advertisement of any proposed purchase of Crown Land. The Land Act of the time allowed staking of government lands, by an agent, for $2.50 to $5.00 per acre. The fee depended on the classification of the land by the surveyor who had been hired to stake the land by the prospective purchaser. Each of the two newspapers often ran five to eight pages of small ads like this one:

FORT GEORGE LAND DISTRICT

Take notice that Anne Findlay of Dublin, Ireland, widow, intends to apply for permission to purchase the following described lands: Commencing at a post planted about four miles north about ½ mile west from north east corner of Lot 1025, thence north eighty chains; thence east 80 chains; thence south 80 chains; thence west 80 chains; to the point of commencement, containing 640 acres more or less. August 3, 1909

Anne Findlay
J.C. Gillespie, Agent

This ad from the September 10, 1910, *Herald* is next to ads differing only in the described location that had been put in for clergymen in England, miners from Vancouver, an accountant in Ottawa, a Cranbrook housewife, a storekeeper from Battleford, Saskatchewan, a farmer from Sandy Bay, Manitoba—and dozens of others. Most of the buyers never saw their land. And never wanted to. When the boom was over the ads were replaced with almost equally long ones of lands for sale for back taxes.

But at the time there was an expectation of easy money. Townsite lots sold for $120 to $500. Businessmen moved in with the hope of quick profits before moving on. The British Columbia and Dawson Railway Company was trying to sell stock for a proposed line to the Yukon. And in April 1910 *Western Lumberman* reported:

WASHINGTON AND IDAHO CAPITALISTS
INVADE BRITISH COLUMBIA

Plan New City on Banks of Fraser River —
Secure Title to 70,000 Acres Embracing
Vast Timber Interests —
Saw and Planing Mills and possibly a pulp mill being planned.

This new town was to be east of Fort George, "situated upon the Fraser River, near the point where the Salmon and Willow Rivers enter that stream." The promoter, Herman J. Rossi of Wallace, Idaho, "one of the most prominent capitalists in the American Northwest," promised a railroad to Summit Lake, mills, and the town of Fraser City. When interviewed, Mr. Rossi mentioned in passing that he hoped to attract settlers. He didn't. The town never appeared on a map.

Only the "Georges" were firmly established. Like all frontier towns they came complete with bars and gambling dens. But competition between the newspapers created a bad reputation for both towns. The *Tribune* was financed by Hammond as part of a half-million-dollar effort to get people to Fort George, and the *Herald* tried to get people to South Fort; so each complained about the other's town, with snide comments like this one from the October 22, 1910, *Herald:* "Central Fort George has no barber shop. They don't need one. The lambs are all shorn in the Winch Building, Vancouver, before they come north."

The *Herald* also suggested that the *Tribune* had started a campaign against gambling immediately after the Central Fort paper's editor lost $16. Eventually the escalating war of insults resulted in a libel suit that dragged on for over a year.

The papers also carried news of who was visiting the area.

> *J. McRae, a timber cruiser for the Yellowhead Pass Lumber Company, has been visiting company holdings on the Willow and Goat Rivers and Ptarmigan Creek with Hiram Carney.*
>
> *Jack Maloney, who has been cruising for William C. Listkow, a Winnipeg timber operator, returned from the Grand Canyon of the Fraser accompanied by Henry Couture.*
>
> *Mr. Russel Peden entertained a party of visitors at his sawmill early in the week driving them into town in his car. Mr. Peden's mill is well situated and is now turning out quantities of splendid fir lumber, both rough and dressed.*

The year 1910 ended on a positive note. Twenty miles east of town, 105 sections on the Willow River were sold for $1.5 million to English buyers, who bought an additional 10 sections at Giscome portage. They immediately announced, in the December 3 *Herald,* that they would erect sawmills in the near future.

Chapter 2
Interlude: From Speculation to Conjecture—The First Royal Commission

In 1909 the B.C. forest industry was out of hand. In the legislature, the Opposition claimed that British Columbia would soon run out of timber. Conservationists looked at what happened in Maine and Michigan and demanded that the government stop history from repeating itself. News stories implied that the government was for sale, just like the forest.

In response, Premier McBride named a Royal Commission on Timber and Forestry headed by Fred J. Fulton, who had been Minister of Lands and responsible for forests until he quit over McBride's handling of railroad contracts in 1908. In 1910 the Commission gave its final report, the most important document in shaping the forest industry in the province.

Sixty-nine years later another Royal Commissioner on Forestry, Peter Pearse, said:

> *I was very impressed with the* **Fulton Report.** *They only had snippets of information to base their report on—what a trapper said he saw—but they drew judicious decisions. They realized how little they knew and advocated a prudent policy that would keep open the government's options in the future.*

The Commission couldn't attack the government, since Fulton had made many of the policies the Commission was to examine. It couldn't attack the lumbermen, since the government was trying to attract investment in the province. To ask the government to break existing contracts would have been foolish, since this would lead to distrust of the government by investors and by the public as a whole. Yet the Commission was under pressure to take action, and the members wanted to. But before problems could be solved, they had to be defined. How much timber had been given away? What remained? Was it any good? What were the best uses for wild lands? What could realistically be proposed?

In less than 100 pages the Commission answered the basic questions and defined rules for government and industry so well that they lasted almost unchanged until after World War II. And when major change came in 1948, it followed a suggestion the *Fulton Report* had made.

The Commission also accomplished something else, something that wasn't required: it made the report easy to understand.

The Commission estimated there were 15 million acres of usable forest in British Columbia, with 3.75 million under government control and the rest in the hands of industry. Fulton, and M. Allerdale Grainger, who wrote much of the report, went on to say that the days of pillaging the forest had to end. In an age when graft and "sharp dealing" were considered just good business, they showed a lot of guts by saying:

> *At the outset your Commissioners were met by the argument that logging operations in this Province were in the hands of practical businessmen...that the forest was receiving the best possible treatment that was commercially possible; that it was absurd to suppose that any timber that could be profitably removed would be left to burn or rot upon cut over lands; and that any arbitrary regulations by the Government would be improper interference with the natural trade conditions of the lumbering industry....*
>
> *Everywhere, in the early developments of lumbering, cheap stumpage is seen to have been accompanied by butchery of wood; for human nature is careless of anything of low commercial value, especially when the supply seems inexhaustible and waste costs nothing to the waster. Where stumpage has risen high in value this state of things has come to an end automatically. Elsewhere a simple legislative remedy has been applied. Operators have been asked to pay for the merchantable timber operated on; not merely for such portion of it as their workmen might choose to cull....*
>
> *Your Commissioners after most careful examination of the theory that unrestricted liberty to allow workmen to cull, destroy, and place in jeopardy of fire the Crown timber of the Province is a fundamental element of profit in British Columbian lumbering, have come to the unanimous conclusion that it is unsound. We are of the opinion that the time is now opportune for the enactment of regulations that will prevent the misuse of the public estate....*

This was, from the lumberman's viewpoint, downright socialism! The man often credited with shaping the report, M. Allerdale Grainger, probably didn't care what the lumbermen thought. A Cambridge scholar, placer miner, ju-jitsu expert, logger, and author of one of the first novels about the industry—*Woodsmen of the West*—he drew conclusions and had attitudes that were straightforward and all his own. The man who became the province's second Chief Forester in 1916, who dedicated his novel "To My Creditors, Affectionately," Grainger remained until his death in 1941 outspoken and uncompromising in his desire to improve the care British Columbia's forests received.

Grainger and Fulton said in the Report of the Royal Commissioners that the object of government policy should be to prevent misuse of forest lands by operators, to "provide for the future of the lumbering industry, secure to the Provincial Treasury revenue that now goes needlessly to waste, and give protection from fire...." They firmly stated that the government, not the industry, should determine which trees should be cut. They wanted payment for wasting the resource, competitive bidding for timber rights, and a provincial Forest Service to police the industry and make sure regulations were followed. And, despite cries from operators that they were going to be forced to hitchhike to the poorhouse, they kept intact the idea that government should share in any increase in the value of timberlands.

However, the Commissioners realized that the solutions they proposed were based on guesses. In discussions of the amount of timber in the province, words like "conjecture," "estimate," "suggest," and the phrase "leads us to believe" appeared time and again. Their attempt to end speculation was based on conjecture. So they borrowed ideas from other countries. The proposal to create a Forest Service was based on both European and American experience. From Germany came the suggestion that the province might want to follow that country's seventy-five-year lead in sustained yield forestry, which replaces cut trees with new trees so that the same number can be cut every year. Sustained yield forestry started in the 1600s with the idea that a tree should be planted for every one cut. Gradually it became clear that in some areas it was necessary to plant only thirty percent to sixty percent of the number of trees desired, leaving the rest to nature. In Germany, sustained yield forestry was almost a science, but in British Columbia the idea did not take root until thirty-seven years after the Royal Commission made its report.

The rules the Commissioners came up with were designed to compensate for their lack of knowledge and to increase the information available. Areas to be bid on were to be surveyed, or, in their terms, "cruised and classified," then bid on competitively. This was to make sure the government knew what it was selling. Provisions were to be made for minor adjustments in the boundaries of land already licenced as knowledge became available. The Forest Service was to find out what was "out there."

Legislation based on the report was introduced two years later by Fulton's successor, William R. Ross. He used words only a politician could say with a straight face:

> *The Provincial Administration of 1905 nailed its colors to the mast: its motto was "public ownership of the forests." Yet it saw that...it was impossible within a reasonable number of years to organize a service that would cruise and value Crown timber, area by area and sell it....*
>
> *The Government threw open all Crown timber lands and anyone was allowed to secure the right to cut timber... the lumbermen of the continent were encouraged to come and operate in the Province, for they could get here all the stumpage they needed. The investors of America—aye, and of Europe—were offered investments. The Government merely handed over the cutting rights, giving an option on timber that was only to be paid for when it should be cut. In return for the cutting privileges granted them, licencees were only required to pay annual interest on the value of the timber reserved for them....*
>
> *In this way 15,000 square miles or over nine and a half million acres were taken up, and the forest revenue doubled and tripled itself....*

The first line of this, the part about "public ownership of the forests," is often quoted. The rest isn't. And Ross had glossed over some things, like why it had allegedly been impossible to organize a Forest Service in 1905 but was simple seven years later, or that the operators would have to pay for the surveys of lands they wanted auctioned off. Instead he said:

> *...An epoch, sir, is drawing to a close—the epoch of reckless devastation of the natural resources with which we, the people of this fair young Province, have been endowed by Providence—those magnificent resources of which the*

members of this Government and this Assembly are but the temporary trustees. That rugged, rudimentary phase of pioneer activity is doomed to an end. The writing is on the wall; the writing—to put the simple fact—is in this Forest Bill. Armed with that weapon, forged by this honourable Assembly, the Government of British Columbia will undertake the work of forest conservation....

Ross was so proud of his speech that he had it reprinted as a booklet. Unfortunately, it had little to do with reality. The rules Fulton had envisioned for minor boundary adjustments were used to remove land from government reserves as fast as applications could be processed. British Columbia lagged behind all other provinces in re-investing forest revenues in the trees that were the source of the money, and continued to lag behind for many decades. The new Forest Act helped. But what really allowed the industry to survive was the combination of the act and the discovery of new lands.

Fulton had estimated that there were just over fifteen million acres of usable forest land. Eight years later a Federal Government survey estimated that there were ninety-six million acres of forest, good and bad. In 1937 a new estimate lowered the number to sixty-five million acres of provincial forest. The number was revised so often that the Forest Service gave up announcing the amount of usable timber in the province. New technology, new markets, or new mapping techniques kept changing things. Trees that were just junk one year would become valuable the next year. So the Forest Service just lumped together everything that had a trunk but wasn't an elephant and classified it as forest.

As the numbers changed, both the government and the operators came to believe that there was a lot more forest than was known. As more and more was discovered, they came to believe that the forest was indeed inexhaustible. As recently as 1976 an operator came to the Forest Service and said, "Well, I cut all the wood there; give me some more. After all, that's your job, isn't it?"

Chapter 3
Boom to Bust
1911-1914

Walf's Sawmill, Summit Lake, 1914. Photo courtesy of B.C. Ministry of Forests.

The government paused for two years after the final report of the Royal Commission before bringing in a new Forest Act. This helped keep alive the boom that might otherwise have died. Speculators rushed in, each sure that the rules of the game were about to be changed. As soon as timber leases stopped being given out, it would mean, they felt, instant wealth for anyone who had obtained a lease in time. This delay also gave members of the government and of the commission time to spread the word, so the government wouldn't appear to be acting hastily. In October 1911, for example, A.C. Flumerfelt—a commission member—wrote in *B.C. Magazine*:

> ...We became convinced that conservation in British Columbia ought to be a very different and very business-like affair. This is what conservation means at the bottom: **the application of ordinary business principles to natural resources....**

But to the population of British Columbia, which had doubled to 392,000 in just ten years, those principles weren't ordinary. "Ordinary" business principles included the ones the land and timber booms were thriving on. New towns were created on paper and made famous overnight. Birmingham, Bella Vista, and Nechako Heights were advertised in what is now Prince George. Nechako Heights still hasn't been developed. Bella Vista slid into the river. The Natural Resources Security Company's ads showed a spider web of rail lines leading from Prince George to Alaska, Vancouver, the Yukon, southeast to Spokane, Washington, northeast to the Peace River Block, and some going nowhere at all. But the ad said these railroads were real, "all chartered, some building." The Grand Trunk Pacific and the Pacific and Great Eastern lived up to this billing, but others, like the Vancouver, Westminster, Northern and Yukon Railway and the Cassiar Central Railway, were merely chartered to give overseas investors something to buy.

When a real railway made a real change—which happened in 1911 when the Grand Trunk Pacific and the provincial government finally agreed about the details of the route across northern British Columbia—stocks shot up. And people rushed to stake more timber. Immediately the few areas that hadn't been staked in the earlier rush in the Upper Fraser and Skeena watersheds were taken up. Land and timber rights were sold and resold by people who had no intention of ever getting within 5,000 miles of the province.

The only people interested in the timber in the woods—not just the paper rights to timber they'd never seen—were American forest interests that had invested over $65 million in British Columbia, and the people in the north itself, like Nick Clarke. And he ran into trouble in March 1911, selling out to a Winnipeg lumber firm.

The "ordinary business principles" of which A.C. Flumerfelt wrote were perhaps best typified by Irene Jordan, one of the unsung pioneer business people in South Fort. According to stories in the *Herald,* Miss Jordan and her friendly female staff were so successful that she was approached by a group of businessmen from Central Fort George to expand operations to their town because of the customers her trade attracted. The businessmen felt that these customers would visit their stores after visiting Miss Jordan's. She did as they suggested. But she built her new whorehouse directly across the street from the church—allegedly because she received additional funds from South Fort George businessmen who wanted to embarrass Central Fort George. Her house was closed down the night she opened for business, so she simply went back to South Fort. In 1918 her establishment was sold to the city and became Prince George's first city hall. The Prince George *Citizen* commented, "If they were to give each councilman his own room, and the mayor the best room in the house, it would be business as usual."

But back in 1911, before Miss Jordan hit the headlines, her main problem was attracting enough girls to keep up with demand. Men were coming north to stake claims. The construction camps became crowded as the line remaining to be built kept shrinking. In March of 1912 rails finally reached the Fraser River just west of Moose Lake. By the end of the year the track was close to Tete Jaune Cache, where two bridges slowed construction. Cooke and Peden's Northern Lumber and Mercantile Company was operating at full capacity, producing lumber for the camps and timbers for the tunnels; the Fort George Trading and Lumber Company, which Nick Clarke had started,

At Finlay Junction in 1914, two men could whipsaw 150 running feet per day. Photo courtesy of B.C. Forest Service.

Cordwood at Fort Fraser, 1913. Photo courtesy of B.C. Forest Service.

Tie making, 1913. Photo courtesy of B.C. Forest Service.

was very busy under the managerial guidance of Mr. C.E. McElroy; and Bogue and Brown's mill was running just as fast as the others. The fourth mill in the north, Tyre Brothers of Tete Jaune Cache, got a contract to provide 600,000 feet of heavy timbers for a rail tunnel near the mouth of the Clearwater River. But the railway needed more than ties. It needed timbers for bridges, poles for stringing telegraph wires, boards for snow sheds and buildings.

Most of the tie camps were just clearings in the woods with anywhere from two to a dozen men keeping each other company, each cutting his own ties. A canvas tent and an air-tight heater, a common washbasin, and nails in the trees to hang things on, a cache on a platform, and all the mosquito netting the men could afford—this was the usual camp.

It was a rough life. Four men died in one accident, two in another. Eventually lost lives became common enough to receive casual mentions in the back pages of the *Herald*. The newspaper gave more prominence to a report that six bears were killed for meat at one camp in a month.

The men lived on hope, 28 cents for each tie, and a lot of potatoes. But they were sure they were where the money could be made. And not just small potatoes. The first pulp mill in the province started operation in April 1912. The new Forest Act finally passed. It was talked about in the tie camps and then ignored until the fall, when Forest Service crews came by surveying along the rail line. The railroad was coming. On July 26, 1913, the *Herald* reported that a group of people from San Francisco planned to run hydrogen-filled airships between Fort George and Edmonton. The story didn't get the lead because Premier Richard McBride had visited Fort George and announced that the Pacific and Great Eastern Railway would be extended south to Panama and north to Alaska, and that sounded much more reasonable than air travel.

The extension of the railway didn't necessarily mean great things for South Fort or Central Fort. The two towns had been planned so the railroad would have to go by at least one of them, since the rest of the flatland in the area was Indian reserve. The Grand Trunk Pacific, which would be the first railway to arrive, had its own plans, however. From just west of Winnipeg to Prince Rupert, the railway had purchased townsites for $1 to $5 an acre along the rail line; unlike earlier ventures, this one did not give the company land in return for building the line. Eighty-six townsites had been laid out in alphabetical order, going from Atwater to Zelma and then starting with A all over again, and again. Today anyone traveling across Alberta can still see the sequence, with missing letters indicating towns that failed. Others, like Wainwright, are the survivors. The plan had to be changed when the tracks reached British Columbia. Some areas, like "the Georges," were already named. And everyone knew that the company would want to develop a townsite where the rails crossed the Fraser.

But where? Since the two Georges were strategically located in the best spots outside the Indian reserve, each had fought to convince the Grand Trunk that it was best.

The railroad responded by paying the Indians to move to Shelley, fourteen miles upriver, and give the railroad the reserve to use as a townsite. In April 1913 the map of the new town, "Prince George," was published. In August the Indians moved out, and the Transcontinental Townsite Company Limited, authorized agents of the Grand Trunk, moved in, pausing only to give the western townsite the name of Vanderhoof, after the railroad's publicity man, and the one to the east the name of Willow River.

When the auction of lots in Prince George was held in 1913, they brought as much as $10,200—the price lots were selling for in the late 1970s. The difference is that in 1913 the lots were prime commercial land, and in the late 1970s they were homesites. It was obvious that Prince George was going to be *the* town in the area. People put their stores and homes on rollers and moved them from South and Central Fort George into Prince. By early 1914 Prince George had a movie theater, six hotels, and a lot of buildings up, or going up.

The arrival of the Grand Trunk Pacific in Prince George on January 27, 1914, bordered on the absurd. Because of a fight between the provincial government and the railroad on how high the bridges had to be, a temporary bridge had supported the rail-layer as it moved into town. The next day an ice jam swept it away. It took three months for the first train from the east to arrive and until August for freight service to really get started.

It was too late.

On June 28, 1914, the heir to the Austrian throne was shot at Sarajevo. In the coming weeks lumber buying across North America declined because of fear of war. When Canada went to war in September, the population of Prince George dropped below 5,000 and stayed down for more than two decades. The other towns in the north were affected the same way. The frontier was for the young—until they could be given a chance to die elsewhere.

In the Forest Service scrapbook, the caption reads, "Sicanni Indians, Fort Graham, Finlay River, who are responsible for fires in that district." 1914.

Chapter 4
The Years of Hope: 1914-1939

When World War I broke out, the boom ended. B.C. lumber production dropped by forty percent from the 1910 level. Mills in the north that had been built to ship lumber via the new railroad went broke.

The Canadian Army organized a Forestry Corps, which cut down and processed trees in France and England to make the supports that lined trenches and the pit props used in tunneling operations and to meet other military needs. Special recruiting of mill men and loggers cut down the number of men available in the industry in British Columbia.

In 1915, the first year records were kept, the cut from the Prince George District was just over twelve million feet. Sure, 137,000 acres were under licence in a six-mile-wide band along the Fraser River, but little was cut. People who had followed the boom followed the bust, moving south, moving east, moving back to the States. The population of the north dropped, then dropped again, and remained a fraction of its boom-period level until World War II.

In July 1915 the first batch of recruits left for the war. When the roundhouse and shops for the railway were completed later in the year, it was steel that kept the city of Prince George alive. Most buildings that weren't abandoned were moved from Fort George and South Fort George into the new city, including Irene Jordan's house and the *Herald* building. When a fire wiped out the business section in 1916, the type and the printing press from the *Herald* were saved, and a new newspaper was started, the Prince George *Citizen*, "Devoted to the Upbuilding of Prince George & Northern British Columbia."

It took until October of 1916 for the lumber market to begin to recover and for people in Winnipeg and points east to realize that Canada couldn't stop because of the war. In October George McLaughlin of the Northern Lumber and Mercantile Company visited the prairies and came back with enough orders for rough lumber to justify building a new mill. This was the first recorded instance of a northern mill going after business instead of waiting for the business to roll in. A new mill was built at Willow River, and the old mill was moved to Giscome, where eighty men cut twenty million feet a year.

There was hope that this would be just the beginning of solid growth. The Pacific and Great Eastern was laying track north of Clinton at the rate of six miles a day. The railroad was expected to reach the city in 1917. Instead, in 1917 the Pacific and Great Eastern became the Pacific and Great Eastern scandal. Although steel ran from Squamish to Clinton, the PGE had gone

Upper Fraser Lumber Company, Dome Creek, 1915. Photo courtesy of B.C. Ministry of Forests.

Breaking up a log deck for Aleza Lake Mills. Photo courtesy of B.C. Provincial Archives. ↓

George Little's Lumberyard, Vanderhoof.
Photo courtesy of B.C. Provincial Archives.

through $20 million worth of bonds and was operating only from Whytecliffe in West Vancouver to North Vancouver. A select committee investigated and found waste, corruption, and bribery. The Premier refused to answer questions about contributions from the PGE to his party. The committee suggested either selling or abandoning the route, and the government ended up taking it over.

The Grain Growers Grain Company, which later changed its name to United Grain Growers, announced the opening of a $150,000 sawmill that was going to produce twenty-two million board feet a year, making it the biggest mill in the north until it burned down in 1926. This mill gave birth and life to the town of Hutton and was the first lumber mill to keep a blacklist of workers who might want a union.

The Giscome mill that Northern Lumber built went bankrupt almost as soon as it was completed. Roy Spurr opened Penny Spruce Mills. And the provincial government started promoting B.C. lumber by publishing a directory of manufacturers in the province and a directory of Canadian dealers stocking B.C. lumber.

Listing who makes what, and where a consumer can get it, may not sound like a radical promotion scheme, but it worked in 1917, it worked in Europe for Canadian lumber producers in the 1970s, and, as this is written, B.C. remanufacturers who take low-grade lumber and reprocess it into higher grades and specialty products are thinking of putting out a directory. After all, it doesn't matter what you think of a product, or how much you want it, if you don't know where to get it. Suddenly people in Highgate, Ontario, could learn that F.W. Scott, right in town, had western red cedar siding and shingles. In Halifax you could visit the Percy Supply Company and get Douglas fir and cedar flooring, trim, molding, sash, doors, and shingles, without having to wait for a special order to arrive.

Similarly, dealers learned from the manufacturers' list that Bogue and Brown produced fir and spruce, common dimension lumber and boards, interior finish lumber, moldings, factory lumber, veneer, large dimension lumber, and ties. Dealers and wholesalers had a choice, of course. If they preferred, they could contact the Burns Lake Trading and Lumber Company, the Deep Creek Lumber Company, Fort George Trading and Lumber (one of the few firms that could actually produce finished lumber, thanks to its planer), J.A. Fraser in Quesnel, the Porto Rico Lumber Company, J.L. Ruttan and Son in Vanderhoof, Northern Lumber of Prince George, the Salmon River Lumber Company of Prince George, Upper Fraser Lumber in McBride, J.E. Holdcroft or L.A. Johnson in the same town, or Andrew Johnson of Raush Valley. According to the government, those were the only mills. The government was wrong, of course. Then, as now, there were seasonal mills and mills whose output was committed to customers. But this list demonstrates that at the time the north was small potatoes. In 1917, total logs scaled came to just over thirteen million feet. That's the size of mills going bankrupt in 1981 because they're too small.

In July 1918 racism arrived in the north with the *Citizen* headline "Injured Hindu Refused Admittance to Hospital." Injured at the Giscome lumber mill and brought by train to Prince George, the man was brought to the hospital but was refused admission by the matron. The mayor and Dr. Lyon argued with the matron and got the man admitted. Two days later the matron was fired.

Prejudice took an even nastier turn in September. Mennonites wanted to move to Vanderhoof. The *Citizen* and the Kamloops newspaper ran editorials talking of "the Mennonite menace," trying to keep them out. On October 18, the *Citizen* said:

> ...The indignation of the "white settlers" is said to be rising like a tide and dire consequences are threatened if the government allows this band of "slackers" to occupy the choicest land in the province.
>
> The Mennonite is a German pure and simple....Far better that Central B.C. remain a howling wilderness than that it become a colony of Germans.

Luckily, the newspaper was given something better to do with its time as the war wound down.

The same issue of the paper carried another article headlined "Assemblies Forbidden in City; Twenty-two Cases of Influenza." Unluckily, the news was bad: the worst killer flu in history had reached the north. Schools were closed. The Connaught Hotel was turned into a temporary hospital. "From the lumber camps east of the city," the *Citizen* said, "comes word of a number of supposed Spanish flu cases." A month later there were seventy-two cases in town. Thirty people had died—most in lumber camps. As everywhere else in North America at the time, the world stopped and waited to see if death would come. It did: to children, to the aged, to the allegedly healthy. Alberta had 7,000 cases of Spanish flu. In Prince George a woman just recovering from the disease had a child, got pneumonia, and died two days later. Two days after that one of her children, seventeen months old, died from the flu. All people could do was wait for the epidemic to pass, which it did toward the end of November.

In the forest industry that year, business was good—except for the months when it stopped because of the flu—but not good enough to justify new ventures. Following the flu epidemic, there was a shortage of men. Ads in the *Citizen* read:

WANTED 25 BUSHMEN
Wages $5 to $5.50
Kitsum Kallum Timber Co.
Remo, B.C.

UPPER FRASER LUMBER CO.
Dome Creek
Bushmen, teamsters, fallers,
swampers

WOODCUTTERS WANTED
Contracts will be made by the British Columbia Express
Company for cutting steamer cordwood during the coming
winter at different points along the Fraser River between
South Ft. George and Soda Creek.

RED MOUNTAIN LUMBER CO.
Penny, B.C.
Wanted: Bushmen & Millmen

On December 20, the government of Canada took over the Canadian Northern Railway.

When the war ended the Coast profited from the opening up of the British market, receiving an order for seventy million feet of timber and ties for English railways. In the north, where the prairies was the major market, the amount of sawlogs scaled rose from twenty-six million in 1918 to thirty-three million in 1919 and thirty-eight million in 1920 as returning veterans demanded lumber for new homes. Shipments to the United States also began to grow. A Quebec firm wrote to the Board of Trade about the possibility of opening a pulp mill in Prince George in 1919, and there was a lot of talk of pulp. But the major firms that would survive the Depression were in place.

According to the Prince George Board of Trade, in 1919 there were

Water tank for icing roads, Eagle Lake Spruce Mills, 1924. Photo courtesy of B.C. Forest Service.

Skidding logs at Dewey, 1923. Photo courtesy of B.C. Forest Service.

eighteen sawmills in the 146 miles between Prince George and McBride, having capacities ranging from 15,000 to 150,000 per day, "and a further increase is confidently looked for in the future."

The reason for the confidence was, of course, the Pacific and Great Eastern, which had laid twenty-nine miles of track north from Quesnel. But there was no money available to bridge the Cottonwood River, so the railroad sold the track it had bought for the last stretch. Within a few years it was on its way to Japan as scrap steel.

On September 3, 1919, the steamer *BX* was wrecked. In November the Federal Government took over the Grand Trunk Pacific.

Then, on January 9, 1920, the *Citizen* heralded the beginning of a campaign for pulp development with its headline "Vast Pulp Resources Should Be Developed." On February 13, the paper reported:

> *"Prince George will be the Spokane of British Columbia," according to Mr. Leonard, a prominent lumberman of Spokane and Florida, when in the city this week.... He expressed surprise at the lack of pulp development here.*

There was a temporary embargo on pulp and lumber from Canada reaching the States, but the *Citizen* suggested it would be short lived.

The talk of pulp continued throughout the year. In December T.D. Patullo, Minister of Lands, received a letter on Empress Hotel stationery from three gentlemen who called themselves the Fraser River Syndicate.

> *Sir:*
> *Referring to our conversation regarding the establishment of a Pulp and Paper Industry in the vicinity of Prince*

Log dump on Eagle Lake in the winter of 1924. Photo courtesy of B.C. Forest Service.

George, B.C.;
We would be prepared to undertake to erect the following plants:
— A Chemical Pulp Mill of a capacity of not less than 100 tons per day
— A Sawmill of a capacity from 100,000 to 125,000 feet Board Measure per day.
...For the first unit mentioned, namely the installation of a pulp mill to produce 100 tons per day of chemical pulp, a sawmill and the development of water power, river improvements, plant site, dwellings, etc. our capital expenditure would not be less than $3,000,000....

Because of high freight rates and capital expenditures, the firm would need in return:

...pulp licences, sufficient areas of good spruce bottom lands bearing timber suitable for the manufacture of high grade pulp.... We estimate the area required would be not less than 170 square miles....
For the above licences we would agree to pay in addition to the fees now provided for, a stumpage not to exceed 25 cents per cord.

The firm also wanted the right to expropriate land for rights-of-way, the right to take as much water as it wanted from wherever it deemed necessary, and so on.

The Crown liked the idea, although it had reservations and suggested a lot of changes. Land was put aside for the company headed by Robert Tyhurst. Great expectations were raised once again. But there wasn't enough

Hauling on "iced" roads, 1924, for Eagle Lake Spruce Mills. Photo courtesy of B.C. Forest Service.

Waterpower sawmill at Stewart Lake, 1924.
Photo courtesy of B.C. Forest Service.

power available to run the pulp mill. For ten years the Fraser River Syndicate tried to get the project going. It failed.

Meanwhile, although lumber couldn't be exported to the United States, dimension lumber was selling, rough, for $32 per thousand board feet, with planing adding about $6 per thousand feet. All the major mills were busy: United Grain Growers, the biggest in the area; Eagle Lake at Giscome, which had changed hands when the mill went broke and was now back in operation; Upper Fraser Lumber Company; Aleza Lake Mills; Hansard Lake Lumber; Penny Lumber; and Gale and Trick at Hansard Lake. Even the eleven smaller mills were keeping busy.

Then, in 1921, although Wall Street was still high, the farmers in the prairies began feeling the pinch that ended up as the tight squeeze known as the Great Depression. The sawlogs scaled dropped from thirty-eight million feet to twenty-two million. Shelley Sawmills went bankrupt.

The Pacific and Great Eastern continued working, south from Prince George, until just forty miles of track remained to be built. But the government still wasn't willing to pay the cost of bridging the Cottonwood, so building stopped—for over two decades. The railroad came north, but operations ran only as far as Quesnel. The government did, however, start building a new highway from Quesnel to Prince George. The population of the north dropped once again, with Prince George going from 3,000 in 1921 to 2,000 in 1928. And as so often happens everywhere when work gets scarce, prejudice raised its head.

The Prince George *Leader*—a short lived competitor to the *Citizen*—editorialized on September 16, 1921:

A WHITE BRITISH COLUMBIA
This is an old slogan, but one that is needed now more

Fire at Eagle Lake Sawmill, 1925. Photo by Harold Mann, courtesy of Prince George Public Library.

Hewn bridge timbers cut by Cranbrook Sawmills, Otway (now in Prince George), 1927. Photo courtesy of B.C. Forest Service.

than ever before.... The fishing and logging operations are now dominated by the Asiatic.... There is a restaurant operated by a wealthy Chinese company in the Cariboo which advertises the fact that a white chef is employed....

The prairies seemed to recover a bit in 1922 as the amount of sawlogs scaled rose from three million to twenty-five million board feet and then to thirty-seven million in 1923. After the embargo was lifted, shipments to the east and then to the States became possible, and the cut climbed to seventy million in 1924 and 1925, reached almost a hundred million in 1926, and then dropped back to seventy-four million in 1927.

In 1927 the PGE's debts had grown to $78 million. A commission of three independent businessmen suggested, again, that the railway should be sold or closed down, and Premier John Oliver said that the remaining forty miles to Prince George "looked like a long-abandoned beaver dam." The only action taken by the government on Premier McBride's old northern dream was to pass the Defunct Railways Act that wiped out 280 assorted railroad schemes. "Northern Development" was a wail of despair, not the battle cry it would become in the 1950s and 1960s.

But even in areas served by the Pacific and Great Eastern or the Grand Trunk Pacific, times were tough. The only mills that survived besides those on the line east from Prince George were some portable mills at Quesnel and Williams Lake, one mill at Vanderhoof, and one at Terrace that produced 35,000 board feet per shift. From 1927 until 1937 only sixteen firms even tried to start new mills, and most of those were very small operations.

It was a quiet time, even though the cut rose to almost 160 million board feet in 1928. By 1931 it was back down to 23.5 million; it bottomed out at 15 million in 1932 and then recovered a bit to spend the rest of the Depression around the 60-million-foot mark.

Quiet does not have to mean dull, however. The Forest Service's annual management reports for the district show, in detail, exactly what the industry was like back then. Here are excerpts from the 1928 report, for example:

...New planers and various other machinery have been installed in different mills and dry sheds are becoming more common.

...The new small mill, "Chabot and Monteith," opened last year, is now operated by other parties as "The Cariboo Lumber Company."

The Lyle Lumber Company has constructed and is now operating a small new mill at Aleza Lake.

Newlands Sawmills, not operated the past two seasons, is logging this winter and operating in 1929.

Etter & McDougall Lumber Company at Shere is approaching the limit of its cut and is looking for a new location. The company can probably operate at Shere for two, or perhaps three more years.

Foreman Lumber Company at Foreman, has completed its cut at Foreman and the mill is dismantled. This company is looking for another location and experiencing some difficulty in finding it. It will probably locate in 1929 at Prince George or in the near vicinity.

The Cranbrook Sawmills at Otway are completing their operation there this season. They have purchased the Red Mountain Lumber Company at Penny and will operate there as such in 1929....

> McLean Lumber Company at Shelley have built a new mill on the Fraser River at that place and have abandoned their old mill located on a slough at Shelley. The old machinery with some additional new items equipped the mill.
>
> The Cook Lumber Company, operating a small mill of about 10,000 capacity near Willow River, discontinued their operation this year and the mill has been moved out.
>
> Bend Lumber Company, Bend, have built an entirely new mill here which started operating late in the season.
>
> Northland Spruce Lumber Company mill at Aleza Lake is still in the hands of the receiver and has not operated this year.
>
> One company, the F.G. Thrasher Lumber Company, finds itself in financial difficulties aggravated no doubt by a serious loss of logs during the period of very high water in the early summer and has called a meeting of Creditors for December 15th. Unless an assignment follows this meeting, there will have been no mill failures in the District in 1928....
>
> An interesting experiment, an expensive one if it does not develop as anticipated, has been undertaken by the F.G. Thrasher Lumber Company in the installation of a dry kiln. The opinion has obtained amongst the mill men in the district, apparently based in some cases at least on careful observation, that our Spruce can not be dried successfully by artificial means....

Later Prince George would have more dry kilns than any other part of the province. At the time, however, working in a mill was a part-time job for farmers from the prairies during the off-season, a 130-day-a-year "full-time" job for loggers. The next year the F.G. Thrasher Lumber mill burned down, kiln and all, Cariboo Lumber at Newlands closed down, and Foreman Lumber started to rebuild its mill—but stopped with the work half done. Jobs became hard to find.

Red Mountain Lumber Company, Penny, B.C., 1926. Photo courtesy of B.C. Forest Service.

Twenty-three fir logs, "scaling 4,630 feet," at Cranbrook Sawmills, 1927

Under the insurance rules of the time, if a mill burned down, it had to be rebuilt on precisely the same site. This rule was instated to avoid deliberate or "fortuitous" fires when a mill finished its cut in an area. During the thirties, suspect fires became so common that the insurance companies paid only when mills were actually rebuilt.

From 1930 to 1939, far more firms failed than started. Often this simply meant a change in management. In 1931 Roy Spurr, his wife, and two friends picked up Eagle Lake Sawmill after Winton Lumber got into trouble. The mill had been changed by previous owners and changed again under Spurr's direction. It was soon handling six million board feet a year with thirty-five men in the crew. Spurr was one of the first millowners to start replacing men with machines, to build all-weather logging roads, and to put logging trucks on timetables. He was the first to supply showers and electric lights for the men in the logging camps in the north, and one of the first to sell mill-built homes to workers for their families.

Today most of those homes are gone. All that's left is a straggle of houses around the lake and the concrete foundations of the mill. The bootlegger who used to come down from Aleza Lake every payday has been forgotten, along with the retired people who often can't afford to live elsewhere. And they're the only ones left to remember the way it was.

Chapter 5
Frustration: The Story of the Forest Service

It seems such a tremendous waste of talent to bog management foresters down with a tremendous load of paperwork.... The Rangers have raised the question of the need for an annual report as its compilation cuts into their time filled work at a time when their work load is heaviest (sic). In this I agree and I again question the value of the written section of this annual report....
Annual Management Report, 1961
Prince George Forest District

Building a lookout tower at Isle Pierre, 1927. Photo by C.D. Orchard, courtesy of B.C. Forest Service.

The author was wrong. Many firms consider everything confidential, and in the sixties someone who was burning a company's old records in a warehouse burned all the Northern Interior Lumbermen's Association records by accident. Companies that go under pitch their records into the trash. Without the Forest Service annual reports, it would be impossible to know what happened. In addition, because the reports were written for the boss instead of the public, they offer a candid perspective on what was happening in the north from a viewpoint quite different from that of the lumbermen or the unions.

When the Forest Service was formed in February 1912, it had a mess on its hands. The land rush of 1907-1912 had to be sorted out. Within six months crews were surveying along the Fraser in the north to sort out timber claims. Two years later the first timber auction was held in the north, in Prince George. Because of lack of staff and money, and a resultant lack of knowledge of the forests, the service was scrambling to do what had to be done at the moment, with little time to deal with things it felt should be done.

The situation remained the same for decades.

In December 1913 the Forest Service had a total permanent staff of 154 employees. Half operated out of Victoria, and the rest were spread throughout the ten forest districts in British Columbia. That works out to 7.7 men to administer an area bigger than New York, New Jersey, Connecticut, and Massachusetts combined. Fourteen years later the staff had doubled to seven rangers, an acting ranger, two clerks, a steno clerk, a stenographer, a chief clerk, an assistant forester, and the district forester. But the district was just as large: bigger than the whole of Ireland. As a result, the annual report for 1927 sounds as happy as a man with both arms in casts and mosquito bites in need of scratching.

> *Ambitious plans for Forest Reserves have been outlined in the annual reports of previous years from this District, but very little has been actually accomplished to date.... There are undoubtedly very great areas in the District which might be reserved as Provincial Forests.... We have no information on most of them and reconnaissance would be expensive. They get practically no protection at this time and are so inaccessible that the cost of protective measures would be prohibitive....*

Forest Service staff from Cariboo, Kamloops, Interior Rupert, and Prince George Forest Districts, 1928. Photo courtesy of B.C. Forest Service.

The report is chatty. It tells how difficult it is to work with the peeled railroad ties the CNR had suddenly insisted on and mentions that the railroad had tried to talk loggers into providing cedar ties, "but the idea is new in the district, and the wood worker, in some respects the most conservative of men, did not appear to be much interested." After discussing each of the eighteen mills in the district, the report gives details of one of the problems the staff was having.

> *...we have kept records on 50 sales awarded during the past six months to determine if possible wherein the waste time if any occurred.... It has required an average of 71 days per sale to deal with these applications. The shortest elapsed period between date application was received and date contracts were forwarded for signature was 16 days, the longest 282....*

Comments like that may be the reason British Columbia doesn't have a freedom of information act. No government wants to air its ineptitude in public:

> *Dishonest scaling under our present system is extremely hard to detect or prove. Both parties have everything to lose by any disclosure and nothing to gain. One scaler did tell me he had great difficulty in obtaining bills for goods obtained from the Company's stores and in making them accept payment. Another that he was strongly encouraged to come late each day to work where hours lost meant logs passed without scaling. One mill was detected sawing logs belonging to another without scaling them and the scaler professing ignorance of his duty. One mill was detected sawing unscaled logs and the scaler signing a fictitious return....*

The report ends with the annual plea for more men.

Though the annual reports are our best record of what happened when, they're incomplete too. Some years are missing. Some reports refer to events everyone at the time knew about—but no one knows about today. And the continual pleading for more money and more men has never been heard by the public.

The Forest Service was originally given the job of protecting and allotting the timber of British Columbia. As the years passed it was also given the job of making sure that the rules of the game were followed. Then it was given the job of bringing the province's forests to a sustained yield basis, so that the amount of trees cut per year was replaced by nature or man and British Columbia would never run out of forest.

The Forest Service failed. In 1981 the province was not on a sustained yield basis. The Forest Service was trying to figure out the rules of the game as they existed in the real world at the same time that it was modifying the regulations. But this wasn't the Forest Service's fault. The reason for this failure is spelled out on page 336a of the proceedings of the 15th B.C. Natural Resources Conference, *Inventory of the Natural Resources of British Columbia 1964:*

Excerpts from "British Columbia Forest Service Expenditure (by annual votes) vs. Direct Revenue Collected"

Fiscal Year	Total Expenditure	Direct Forest Revenue	Expenditure as Percent of Revenue
1912-13	$268,163	$2,569,003	10.4%
1915-16	356,008	1,709,657	20.8
1920-21	577,288	3,315,891	17.4
1925-26	1,045,099	3,592,899	29.1
1930-31	1,026,177	2,914,638	35.3
1935-36	697,698	2,841,419	24.5
1940-41	1,051,805	3,549,932	29.6
1945-46	1,599,181	4,352,179	36.7
1950-51	4,438,506	10,089,855	44.0
1955-56	8,159,977	23,867,706	34.2
1960-61	18,826,015	29,819,295	63.1

Then the percentages began dropping. According to the F.L.C. Reed & Associates report, *Forest Management Expenditures in Canada,* in 1979 the province received $565.6 million but spent only $210.3 million, or thirty-seven percent, on forest management. It's no wonder that throughout every annual management report there's a feeling of frustration, of being forgotten, of the boss not caring. Invariably, the reports are written by someone with a passionate belief that what he and his staff are doing is important but who feels unable to do the job that has to be done because of lack of staff, lack of funding, and lack of direction from Victoria. An attitude of near rebellion is

Cruiser's camp, 1915. Photo courtesy of B.C. Forest Service.

Fire wardens the year of the new Forest Act, 1912. Photo courtesy of B.C. Forest Service.

Finlay Junction Ranger Station headquarters, 1914. Photo courtesy of B.C. Forest Service.

common. The men in the field usually felt that the people in Victoria had no idea what they were talking about, whether the subject was logging regulations to be enforced with eight feet of snow on the ground, as is common in the north and not thought about in Victoria, or the need for equipment to cope with literally trackless wilderness.

C.D. Orchard was the ranger at Aleza Lake in the late 1920s and ended up as the Chief Forester in the 1940s. He felt so strongly about the bleeding of the forest that he wrote a pamphlet on the subject, which came out in 1949 and was revised twice. Here's how he put the numbers in the 1953 edition:

> ...If we had been wise enough to have introduced effective forest management at the time (1900),*we could have ensured a "perpetual" minimum annual harvest of perhaps 1,000 million cubic feet, which could have been increased... to 2,000 million cubic feet....
>
> If our fathers had instituted such a policy at the time, British Columbia today could be cutting an estimated 1,300 million cubic feet, or a **safe** 50 per cent more than our actual present cut of 880 million cubic feet which is bordering on the danger line....
>
> Our bank account is being withdrawn in increasingly large amounts.... Retrenchment, whether it be in the family budget, the national budget, or the forest budget, is always a painful process; but to balance income against expenditure, growth against harvest, is the only way to keep out of the poorhouse.

The government ignored this advice, given six years after the province was put on an allegedly sustained yield basis, and the government has continued to ignore it ever since.

The reaction by the Forest Service was predictable. To the public, and to the industry, the Service desperately strove to give the impression that it knew what it was doing and that the situation was under control. But its internal papers sounded far less confident—and far more like a cry for help.

* Unless otherwise indicated, in all quotes information in parentheses is author's addition.

District Forester's headquarters at Tete Jaune Cache, 1913. Photo courtesy of B.C. Forest Service.

Chapter 6
How Big Is the Forest?

Hauling for Sinclair Spruce Mills. Photo courtesy of B.C. Provincial Archives.

As soon as the Forest Service was established in February 1912, it went to work mapping the province's forests, trying to find out what was where.

In 1907 Dr. Bernard Fernow had said the earlier estimates of over 180 million acres of forest were far too high and suggested that the real number was between 30 and 50 million acres. Then the 1910 Royal Commission estimated that Fernow was wrong and that there were really only about 15 million acres outside the Dominion Railway Belt. But the Commission didn't really know.

The Forest Service began its work, logically, in areas where there was a lot of action. In the north this meant along the Fraser River and its tributaries. By 1914 the areas people were busy claiming had been looked over. But this left over ninety-five percent of the north still unknown, and the Forest Service had to rely on maps made by Father Morice that were based on what trappers and Indians said the land looked like. As a broad outline, the maps were surprisingly accurate, but for working out how many acres of each species grew, and how good the trees were, the maps were almost useless. An area of 500 square miles might have the simple notation "pine and spruce."

In 1918 the Federal Government tried to gather all available information on how big the forest was. Its estimate was 96 million acres.

In 1937 the first detailed study, by F.D. Mulholland, finally changed the practice of lumping together statistics for the province. He found that considering accessible forest only, the Fort George District had a sustained yield capacity of 185 million board feet per year, the sawmills cut 66 million board feet, and fires destroyed 13 million board feet. His report told more than statistics, although it had plenty of those, of course. It reflected the times. Here's his description of the area:

> The most northerly of the organized districts, in the north-east and east-central part of the Province, its principal drainage basins are the Upper Fraser River and the headwaters of the Peace River and its tributaries, the Finlay, Parsnip, and Omineca....
>
> Forest utilization is confined to the southern part of the district, served by the Canadian National Railway, which follows the Fraser and Nechako Rivers on its way from the East to Prince Rupert, and the Pacific Great Eastern with its terminus at Quesnel on the southern boundary of the district.

> *The timber is the northern forest type of Englemann spruce, silver fir, and lodgepole pine, with, in the south, some Douglas Fir.... Two-thirds of the merchantable volume is spruce.*
>
> *Woods operations are carried on both winter and summer; logging is still chiefly done with horses, but caterpillar tractors and trucks are being introduced.... There are sixty-one mills, with a daily capacity of 740,000 F.B.M. of which forty-eight (capacity 586,000 F.B.M.) were operating in 1936. Most of the mills are small, only 15 having a daily capacity of over 20,000 F.B.M. and two over 50,000 F.B.M....*
>
> *The sawmills are so far from their very competitive markets that they are perpetually fighting the handicap of long rail-haul and freight charges. It is unlikely that any great increase in utilization will occur until economic conditions affecting the world pulp market are such that it will be possible to construct a pulp-mill to utilize some of the spruce, or until greatly increased demand develops on the Prairies....*
>
> *The immediate value of the accessible forest is very great; they provide the only industrial pay-roll in the district, excepting the railway itself and the construction and maintenance of Government roads. Tie-making and other work in the woods and mills are the means of existence for many of the settlers.*

The Forest Service continued trying to assess the forest it managed, but because of lack of funds progress was slow. In 1945, when Justice Sloan was working on his report, there were still gaping holes in the inventory. Mulholland had at least left some detailed information for Sloan to work with: 17,000 miles of examination strips on the ground and 3,000 miles of aerial photography. However, as Sloan commented in 1956,

> *This inventory was made during the depression of the thirties, and its assessment of the qualifications "merchantable" and "accessible" were coloured by the atmosphere of poor markets and low profits.... Many stands which were correctly mapped and compiled at that time as inaccessible...or as worthless scrub have since been logged.*

Survey worked halted by the war was started up again in 1945. The Cottonwood, Naver, Crooked River, and Stewart Lake areas, surveyed as part of the 132,708,445 acres of the province, had been looked at by 1956.

This may sound as though the province knew by 1956 exactly what was in its forests. It didn't.

The start of each survey was aerial photography, followed up by sampling on the ground. In 1956 the photos of some areas still hadn't been transferred to maps. Other areas had been looked at briefly, with growth factors worked in automatically and no account taken of damage by pests, windfall, or weather that affected growth. Soil types that determined rates of growth weren't known for all areas. In his 1956 report, Justice Sloan included the following dialogue with Mr. Pogue, who was responsible for much of the survey work:

Photo courtesy B.C. Provincial archives

Q. But today, do we know what we have in the timber bank?
A. It is obvious that we don't.
Q. And, save for a complete analysis, based on a ground cruise, of the whole province, is there any possible way in which we could find out what we have in the timber bank today?
A. We are doing a Provincial survey as fast as we can. It is a sort of 5 cents an acre coverage basis. Now, possibly the type of answer which is being hoped for won't come out at anything less than a $2 an acre survey. If there are a hundred million acres of productive forest land, even if we are able to concentrate on them, that is quite a tall order.

It was more than a tall order. It was an order that the Forest Service felt, for a long time, was designed to make it look foolish.

According to one Forest Service member:

> *Every time we'd work out the numbers, a species we'd considered junk would be developed for an economic use, like the time Leon Koerner changed the name of hemlock to Alaskan Pine so it would sell in England. Or else a new level of technology would develop, like when Caterpillars really got going, or high-lead logging. And bingo, we'd have statistics wrong the day they came off the press. Finally we said, "the hell with it" and lumped the forest together with the species shown and let others decide what was "merchantable," what they could sell.*

The following numbers suggest that British Columbia shrinks and stretches like a rubber band. Here's the amount of timber in British Columbia, according to the experts of the year involved, up to 1956:

Year	Million Board Feet
1910	240,000
1917	336,300
1937	254,500
1945	303,300
1951	431,500
1955	760,050

This wasn't the end of the pulsations in size of the B.C. forest, of course. But already people were getting some ideas firmly fixed in their heads.

Industry became convinced that the Forest Service consistently underestimated the size of the forest. And of course it did. The Hanzlick formula, used to work out how much could be cut each year on a sustained yield basis, assumed that trees stopped growing when they should be cut under ideal conditions. That underestimated the amount of wood. And since foresters knew that their data were crude, they cut down the numbers in their sample plots. As one forester said, "If a logger only got out what the Forest Service said was there, he was creaming the stand taking only the best timber. I remember when you guys (in the industry) would take out a third more."

And though the Forest Service and the industry knew that all the numbers used were estimates, the estimates were stated precisely, so the public thought the Forest Service had the situation under control.

In the 1964 *Inventory of the Natural Resources of British Columbia*, the proceedings of the 15th B.C. Natural Resources Conference, there are neat tables showing that Prince George had nineteen Public Sustained Yield Units (PSYUs) containing 9,674,758 acres of sawtimber. This includes the Canoe PSYU running south from Tete Jaune Cache along the Alberta border. But according to the Inventory Officer for the area, the first survey of the Canoe PSYU that took more than a glance at the area was done in 1974.

Hauling near Sinclair Mills, using logs for rails. Photo courtesy of B.C. Provincial Archives.

For decades, the Forest Service spoke in precise tones about areas where it knew very little, drawing fine distinctions on barely more than guesswork. Because of the high cost of good surveys, it had no choice but to guess. The Service, however, has always been reluctant to admit how much of its work is based on what it hopes is intelligent conjecture. Again, the Forest Service had no choice. The officials could hardly be expected to get up and say, "We don't know what we're doing, the government you folks elected is a failure and is putting far too little back into the forest, and the biggest industry in the province should shut down while we find out what's going on." If anyone said that, he'd soon be transferred to Atlin or Alice Arm—and have no chance of doing what was possible on limited funds. The role of the Forest Service has always been to try to control and manage the forest on a limited budget.

Luckily for the province's future, the numbers have been getting better and better. By 1976, Peter Pearse was able to write in his Royal Commission report:

> *The Inventory Division of the Forest Service has developed a sophisticated forest inventory programme, based on field sampling and photogrammetric techniques. Continuous surveys result in periodically revised estimates of the forest cover and growth rates in each management area....*

In 1976 the acres of forest in the province were estimated at 128,748,600, up from 90,527,809 in the 1962 estimate.

The number was wrong, of course—at least according to the 1980 estimate. But at least it was closer than previous numbers.

In 1981, Bill Young, the province's Chief Forester, said:

> *At last we know what we don't know, and have a reasonable way of filling in the gaps. Survey and inventory work has been accelerated, and our numbers are better than ever. By 1990, we should be able to tell, at last, how much of what species is where, and have it right.*

The point is simple: no one really knows, or ever knew, what the province's forests contain. But the industry, the Forest Service, and the politicians have always acted as though they did.

Chapter 7
A Hell of a Way to Earn a Living

Bucking logs to length at Trick Lumber, Aleza Lake, 1935. Photo courtesy of B.C. Forest Service

It could have been 1924, 1936, 1943, or, in some camps, even 1950. The camps didn't change. The tick that passed as a mattress was filled with fresh straw in the fall and not changed until the next summer. Each man had to bring his own blankets. The single seat in the bunkhouse was a pants-polished slab under a kerosene light high on a roof beam. At one end of the room a wood stove was surrounded by a spider's web of wires and rods for drying clothes that were washed whenever people complained about the smell. A bunkhouse at the usual camp until about 1939 was as dark as a bear's cave and smelled as though the bear was wet. It had all the frills of an axe handle.

In the second slab-roofed building the ink-slinger—as the bookkeeper was known—the woods foreman, and one or two others would have more room but similar accommodations. Often there would be a combination office, store, and storeroom at one end of their building.

The work schedule was simple. Log in the winter, mill in the summer, and survive on your own during the break-up and freeze-up.

In the woods, work started when it was still dark. Up, breakfast, harness the horse, then three men would walk out to the stand of trees they were working to be in place at dawn. One man, sometimes a faller, sometimes no better at falling trees than the others, would cut the trees down. The second man would limb them. The third would use the horse to haul the logs to the edge of water. On the Coast these men would have titles: buckers, sawyers, swampers. In the north these were signs that a man had either come from the Coast recently or else read novels a lot. These nicknames were as hated as the term "lumberjack."

McKinnon's eight-horsepower sawmill on the Finlay River, 1928. Photo courtesy of B.C. Ministry of Forests.

All winter the logs would be piled on the edge of frozen lakes and rivers. At Giscome they'd be hauled on sleighs in apartment-house-sized loads. Elsewhere they'd be dumped into the water in the spring, to float to the mill.

In the summer, the same men would turn the logs they'd cut into rough lumber. In between they'd be in a flophouse in town, hoping they wouldn't drink up their pay before the ground got solid.

In the late 1920s, wages in the north were 45 to 55 cents an hour for sawyers (in the north this term was used only around a mill for the guy who ran the saw), a nickel less for teamsters, camp help, and the guys who cleaned up. A camp foreman made $125 to $150 a month. Railroad ties were paid for by the piece, with the Prince George Forest District annual report for 1928 detailing:

	Per Tie
Making (usually includes stacking convenient to haul road)	$0.14 to $0.20
Skidding when necessary	.0025 to .02
Peeling	.02 to .05
Supervision	Nil to .05
Office	Nil to .01
Workman's Compensation Board	.005 to .008

Then the Depression hit and wages were cut back to 22 cents an hour. Suddenly ideas that had seemed too radical for the north, like unions, seemed a lot more reasonable. Anyone with a job was smart to keep it for as long as possible, so the men stayed in the same camp instead of moving in the spring, the fall, and when the whim hit them. People became attached to their camps and organized debating societies, dances, and movie nights, and even built ski trails.

But working conditions in the mills stayed the same. Some mills were just jacked-up Model T Fords running open saws. Others had steam-powered gang saws. All were crude. A "Special" on the CN to haul a mangled worker to Prince George or McBride wasn't unusual. At one mill, one man a month was killed for six months. Almost every issue of the Prince George *Citizen* reported, "Man Killed by Snag" or "Alex Sandhurst is doing well, we're glad to hear, after his accident." Accident reports of the time suggest that sawmill designers were practicing to become mass murderers.

One type of mill cut piles of logs at a time so that loose slabs of semisawn wood would spew out the far end. Often a slab roof protected the saw but ended just above the man unloading lumber. If it rained, it was like unloading boards in a car wash. At Penny, where the mill was built on stilts on the river bank, a man might lose his leg riding the carriage. According to the people who worked there, a man once ran the carriage off the end of the track, through the wall, and twenty feet down to the ground, with the setter and dogger still aboard.

When the mills closed down in the fall, or when logging was halted by spring mud, men from all over the north came to Prince George. Most people in town stayed away from the loggers. "The whores were the only real friends we had," one logger said. "They'd talk to us even when we were broke. They were outsiders too." In a way it was understandable. Loggers spoke a different language from townsfolk. Some smelled like old gym socks, and they wore spiked boots to dances.

Loggers went through money so fast that one camp operator who also owned a bordello would pay the men in cash; then, before they could leave camp, she'd send over three taxis—two full of girls, one of booze and cigarettes. Telling the story, an old logger said, "Finally one man said

Rolling logs down a skidway at Longworth, 1925. Photo courtesy of B.C. Forest Service.

A "four-up" near Sinclair Mills. Photo courtesy of B.C. Provincial Archives.

A plank road under construction near Upper Fraser. Photo courtesy of B.C. Provincial Archives.

Sinclair Spruce Mills. Photo courtesy of B.C. Provincial Archives.

'Carmen, I get paid in the same bills every payday. Why don't you just give me a credit?'" When he was asked if this story was true, the logger replied, "Hell, I didn't make it out of camp for almost two years!" Another pair of operators had a mill where the Prince George airport is now located. The bark shacks they set up for the men were so bad that the guy on the top bunk could see the stars at night. But no one quit. The boss was a professional cardsharp, and he started a poker game every payday so that none of the men had money to go to Prince George and bunk-up until another job came along.

Talking quietly at the Rainbow Senior Citizen's Home in Prince George, John Hemming made the old days come alive:

> I was near everything, though I never worked in a mill. I was a logger, a flunky at Sinclair Mills, felled trees, built plank roads. I spent a lot of my time at Sinclair Mills building plank roads. We'd use three- or four-inch planks nine or ten feet long: spruce. At Giscome they used fir. Fir was better, stronger. It was a constant job pulling up bad planks and changing them. Though we'd spike everything down, it'd soon get loose. And just one mile would take 1,800 pounds of size seven or eight spikes. The basic road would last until someone drove off it, which was all too common. That would put pressure on the edge and just rip it up. We used what we could get no matter what the width, and we'd use trees for rails underneath—four in a row. That may seem like a long time ago, but you've got to remember that plank roads kept being used around here until past 1950.

Plank roads, steam radio, and 5,000 men scattered in camps across the north; the number dropped to less than 3,000 in the mid-thirties, then went up again as packing cases were needed for shipments to countries in Europe preparing for war. The war started with tanks and ended with the atomic bomb, but the mills and the camps in the north had barely changed since the early 1920s.

They were about to.

Hauling for Aleza Lake Mills before 1920.
Photo courtesy of B.C. Provincial Archives.

Loading truck, one log at a time. Photo courtesy B.C. Forest Service.

Winter at Sinclair Spruce Mills. Photo courtesy B.C. Provincial Archives.

Springtime at Upper Fraser. Photo courtesy B.C. Provincial Archives. ↓

Chapter 8
World War II: Father of the Union

Poland was invaded September 1, 1939. Two days later England declared war on Germany. A week later Canada went to war. But even before bullets were fired, the threat of war had helped stir Canada out of the Depression. A few pages away from the ads for coffee at 25 cents a pound, hamburger at 15 cents, and seven-year-old rye at $2.15 a bottle, the *Citizen* had reported:

> HEAVY INCREASE IN LOG SCALING
>
> *August Figures for 1939 Show Threefold Increase*
>
> *Log scaling in the Fort George forestry district for the month of August this year totalled 12,733,205 board feet as compared with 4,052,908 board feet in the previous year....*

At the end of September, the government halted survey work for a highway to Alaska. The Commonwealth Air Training Plan was announced. It needed 93 airfields, including 100 all-wood hangars, barracks, instructional huts, and maintenance shops. In one season 5,000 new buildings for workers at 160 sites across Canada went up. Then the Commonwealth started sending shiploads of British WAAFs for radio training, Australians to learn maintenance, and tens of thousands of recruits from apprentice spies to ski troops. More housing was needed. Factories had to be rebuilt as existing plants changed products. Montreal Locomotive Works, for instance, began producing tanks. Sitka spruce trees in the Queen Charlotte Islands were cut and shipped to England, where they became Mosquito bombers. Then more Mosquitos were built in Canada. The Pacific was threatened, so more boards were needed to build camps to defend Canada's western flank. Chips became sterile dressings. More and more men left to join the army, including a Forestry Corps with almost 2,000 trained loggers and mill men from British Columbia. But still more lumber was needed.

Harry Neal, who helped develop the IWA, remembered:

> *There was a camp behind every stump in the woods. You could sell a stick that was half bark, and a lot of operators did just that. Some guys would cut trees, roll 'em on a skidway, then go to the bank and say, "Hey, we'll use our trees as collateral—give us the money for a truck." As soon as the truck was paid off, they were back for another loan to buy a saw.*

From 53 million feet in 1939, the log scale went to 72,642,100 in 1940. Timber sales rose from 179 covering 30,571 acres to 239 involving almost 52,000 acres in 1941, then jumped to cover almost 72,000 in 1942 and 77,228 acres in 1943.

Sinclair Spruce re-opened its camps at Longworth. The Guilford Lumber Company took over the old Vick Brothers mill east of Penny. Suddenly there were more than forty new mills of various sizes in the district.

Because of the shortage of men, the operators improved working conditions as well as living conditions. Orders were coming in reading, "Send us a boxcar full of boards. Any kind of boards." Logging was declared an essential industry. The bigger operators at Penny and Sinclair Mills, Upper Fraser and Giscome, Prince George and Quesnel milled day and night as long as there were logs to cut.

The Forest Service was short on staff, like everyone else, with each ranger averaging 243 field inspections in 1940.

The U.S. Army began surveying a different highway route to Alaska. Fire balloons sent by the Japanese started fires as far east as Penny.

"We were a mixed bunch," John Hemming said of Sinclair Mills. "There were Japanese, immigrants from all over Europe, farm boys from the prairies, and even some raised in B.C." At Sinclair there was still a six-hole outhouse, but the food—which had been good—improved. Trailers were hauled in by railroad to ease the strain on the bunkhouse. A new washhouse was built.

The 817,861 residents of British Columbia identified by the census were in the midst of a timber binge. In 1942, 1600 logging operators were supplying 551 sawmills in the province, 4 plywood plants, 76 shingle mills, and 7 pulp and paper mills. Of all the capital invested in British Columbia that year, forty-five percent, or over $154 million, was put to work in the forest industry.

C.D. Orchard, the Chief Forester of the province at the time, wrote a memo to the Premier in 1942 that was prompted by the growth in the industry:

> *British Columbia must formulate a long term management policy. After a lapse of a short twenty-nine years our visible resources have shrunk from two hundred and fifty years' supply to thirty-three years' supply.*

The opposition also noticed the way timberlands were being thrown open. The Co-operative Commonwealth Federation began regularly asking questions in the legislature. Colin Cameron's pamphlet, *Forestry...B.C.'s Devastated Industry,* was selling well.

In 1943, the peak of the boom passed. Although lumber was still shipped to the United States and Europe, the amount produced in the province began dropping toward the end of the year. Since most exports overseas were from the Coast, the number of timber sales in the north showed the change in the North American market quite clearly. From 77,228 acres in 427 sales in 1943, the number dropped to 39,950 acres in 255 sales in 1944, before going back to 50,679 acres in 1945.

These figures were still far higher than before the war, of course. Wages were frozen and men were so scarce that the Forestry Corps was one of the first units returned home at the end of the war. Living conditions in the camps had improved greatly, but as the standard of living went up, the men in the camps began to hope for and expect more improvements.

The hot stove league in the camps debated solutions to the problems of low wages and bad working conditions. Socialists, communists, new worlders, free thinkers, apprentice capitalists, and representatives from every other viewpoint had their say. This wasn't new, but the war brought a new mood:

the bull sessions ended and serious debate began. Wobblies who had been in the International Workers of the World (IWW) in the Pacific Northwest states added their opinion. People who hated unions spoke up. Men who had watched the union movement grow on the Lower Mainland and Vancouver Island told of the good points and the bad. By 1944 the question wasn't whether the workers in the northern mills wanted a union, but which union they wanted. At Sinclair Mills one man with IWW experience got up and said:

> Boys, you don't want the International Workers of the World. What you need is a responsible union that management will respect, instead of making the bosses worry about revolution. Don't get sidetracked trying to change the world when you really just want to change working conditions.

The bush telegraph carried his message to every camp in the north.

Then, in 1945, Don McPhee, who managed Sinclair Mills, and Roy Spurr, the boss at Eagle Lake, tried to get the men a raise. The Labor Relations Board turned them down. Each told the workers it was because they weren't unionized. Both camps invited the International Woodworkers of America to send an organizer.

Just after V-E Day, Ernie Dalskog came to Eagle Lake. In two nights he signed up over ninety percent of the men. By the end of the week every man in the operation was a member. Then Six Mile Lake Sawmill joined. In two nights the staff at Sinclair Mills signed up. On July 28 the new union local, 1-424, had its first meeting and called for a 48-hour work week. By the time the Labor Relations Board was shut down in the aftermath of the war's end, almost 75 operations in the north had been certified. That left only 675 mills in the north out of the union.

Photo courtesy IWA local 1-424.

Logging crew near Prince George, 1952.
Photo courtesy IWA Local 1-424.

Chapter 9
The Second Royal Commission: Political Flim-Flam Strikes Again

For years, when anyone in the Forest Service said "sustained yield," his voice deepened to the tones of a television evangelist. When the forest industry was under attack, it spoke of "the 1945 Royal Commission that brought in our present sustained yield policy" as though an eleventh commandment had been added that year.

Sustained yield *is* a good idea. But British Columbia has never had it.

When the Commission suggested that a tree should grow for every tree harvested so the yield of the forest would be sustained, Germany had already practiced sustained yield forestry for a century. The 1910 Royal Commission had suggested the idea. In 1925 the B.C. government set up Forest Reserves, allegedly for sustained yield forestry, and by 1944 these included over twelve million acres. This wasn't a new idea. The decree setting up the 1945 Royal Commission specifically directed Chief Justice Gordon McFarlane Sloan to investigate "the establishment of forest yield on a continuous production basis in perpetuity."

But in the forests, in the forest reserves, there weren't any signs of sustained yield forestry. The government had marked maps but hadn't provided funding or planning.

The change that made sustained yield an official government policy wasn't a change in ideas, it was a change in politics, and this change created the 1945 Royal Commission.

In 1941 the Co-operative Commonwealth Federation (CCF), predecessor to the New Democratic Party (NDP), had published a 5-cent pamphlet called *Forestry...B.C.'s Devastated Industry, A Frank Discussion by Colin Cameron, M.L.A.* The pamphlet sold out, was reprinted, and sold out again. Everyone in the province could see in the frenetic logging for the war that Cameron was right when he said that the province was "like an exiled Russian princess selling off her jewels one at a time." The government was selling off its forests, he said, instead of investing in them and living off the interest they provided in the form of new growth. The voters agreed with Cameron when he concluded:

> *There is no possible solution to the problem of preserving our major resources and industry unless private ownership of forest lands is abolished or unless the private owners are prepared to operate under the complete control and supervision of public officials in a compulsive and integrated state scheme.*

So, in 1943, the government called for a Royal Commission on Forestry. Although the Commissioner, Gordon Sloan, spent 111 days listening to the manager of colonization for the CNR, the Fort Fraser Farmer's Institute, the Slocan Women's Institute, and dozens of industry groups and government experts, the real action took place when Cameron and Sloan argued. Despite the government's requirement that the Commission look at forestry education, the size and type of forests in the province, and sustained yield, the real question Sloan had to answer was: exactly what did the government have to do to get Cameron to shut up about nationalizing privately-held forest lands and forest companies?

Cameron was stubborn. He wanted exactly what he had written.

And Sloan gave it to him. In spades. He provided exactly what Cameron wanted in a form Cameron had difficulty complaining about, while accomplishing exactly the opposite of what Cameron desired.

The report opened with words Cameron agreed with: "Our forest land must be regarded as the source of renewable forest crops and not as a mine to be exploited and abandoned...." It went on to echo Cameron, saying that "millions and millions of dollars have been drained from our forests." Unlike Cameron, Sloan showed the statistics that proved the point. Depending on how you want to measure income from the forest, reinvestment of revenues had averaged either less than 21 cents per dollar or just under 29 cents per dollar. Either number was less than the amount invested by any other province with the exception of New Brunswick. The U.S. Forest Service had been investing more than it received in direct revenue. How could it afford to do so? According to Sloan:

> *There can be no doubt of this fact: that the multiplicity of forest functions and the total indirect values derived therefrom bring to the public generally, and the Provincial coffers in particular, far greater sums of money than is realized from measurable direct revenue.*

It was obvious, Sloan said, that tourists wouldn't spend millions "if all we had to offer our visitors was the devastation of logged and burned lands and dismantled scenery."

Cameron must have been delighted. And—though the quote was used to imply Cameron had said nothing new—he was undoubtedly pleased when Sloan quoted the 1910 Royal Commission as saying:

> *Your Commissioners regard the income from royalty on timber as differing essentially from any other form of revenue in the Province. Such receipts should be regarded as capital not as current revenue....*

But when Sloan dropped from philosophy to reality, he dropped Cameron. To be kind, Sloan was shortsighted. Although lumber production had tripled between 1915 and 1944, he assumed it would never rise again. Apparently Sloan used Mulholland's 1937 figures on the amount of timber in the province, subtracted what had been cut since Mulholland's estimate, and then subtracted a fudge factor straight from the candy section of the *Fannie Farmer Cookbook*. According to Peter Pearse, Sloan took his recipe for sustained yield forestry from a memo Mulholland had written in the late 1930s, and the secret of his Improved Forest Service Frosting is a mimeographed submission by C.D. Orchard. Here's how Sloan put it together, on pages 127 and 128 of his report:

> *I would define "sustained yield" to mean a perpetual yield of wood of commercially usable quality from regional areas in yearly or periodic quantities of equal or increasing value....*
>
> *...the next step must be a consideration of the various things that must be done to bring about a sustained-yield policy....*
>
> *(1.) Fire-protection must be greatly increased.*
> *(2.) The rate of planting denuded areas of productive forest land—especially on the Coast—must be greatly increased.*
> *(3.) Logging methods must be regulated to prevent destructive exploitation and to ensure full regeneration of cut-over lands.*
> *(4.) New systems of tenure and taxation need to be formulated to encourage private forestry and to remove causes compelling liquidation.*
> *(5.) Management plans for individual regional working circles should be formulated and implemented by regulation....*

Sloan suggested other things: a Forest Commission to oversee forestry management, more education, intensified research; but they weren't acted on.

What his plan boiled down to was simple. If it took trees eighty years to grow to maturity in an area, then obviously one-eightieth could be harvested every year, as long as a new crop got started. Fire protection would prevent loss of the crop. Planting of areas that were logged would ensure that areas grew up with trees rather than brush.

To encourage management of privately-owned land, Sloan suggested "private working circles." These would combine private holdings with some specific public land in the area, all under the control of a forest company that would be held responsible for forest management. When the government adopted this idea it called these areas Forest Management Licences, then changed the name to Tree Farm Licences. To keep things simple, they will be referred to throughout as Tree Farm Licences, or TFLs, as they are popularly known.

The government liked the idea because it put the burden of management on the companies. And, since it was up to the Minister of Forests to decide who got a TFL and who didn't, this power of allocation added to government control over the forest industry. Companies liked the idea because they got additional land that was, for all practical purposes, theirs alone, without having to pay for it. Originally the licences were given on a perpetual basis. Later the licences were modified to twenty-one years with almost automatic renewal. From 1947 until 1976 no firm ever had TFL land taken away for mismanagement, even though 34,642 square miles involving thirty-four TFLs had been given away by then. There was a good reason for this. Since the companies involved were sure they'd get the benefits of any forestry work they did, spending money on forestry made good economic sense. As studies around the world have shown, people tend to work hardest on forest lands under their control.

The Forest Service says, of course, that it oversees forestry and harvesting on TFLs to ensure that guidelines are followed, but in reality companies have a great deal of say about what's done and what's not done on TFLs.

Although applications had to be advertised, the fights for TFLs were sometimes bitter since the legislation put the whole matter under Ministerial discretion.

This wasn't what Cameron had in mind at all.

Sloan suggested that land not included in TFLs should be turned into "public working circles." Again the name changed, first to Public Sustained Yield Units (PSYUs). Then, in 1978, PSYUs were lumped together into Timber Supply Areas. Since most people in the government and industry still think of them as PSYUs, that's the name you'll see for the rest of this book.

Sloan's idea was that the Crown, through the Forest Service, would manage the PSYU lands and say who could cut how many trees, with "the production...to be allocated to the open log market or to small millers." Note the word "allocated." Again it meant that the government would decide who got what—and who got left out in the cold. At first Sloan accepted—and the government went along with—the recommendation of the Northern Interior Lumbermen's Association that "all existing operations should be brought under licence and no new operation should be licenced until the availability of timber in the area affected, sufficient to guarantee the maintenance of that operation on a sustained yield basis in perpetuity, has been determined." This, of course, boiled down to protecting existing operators at the expense of others, and the idea was eventually—at least officially—dropped.

Sloan was a success. He removed the threat to the government of the day by assuring the voters that the province wouldn't run out of trees.

It wasn't what Cameron had planned, of course. Cameron had envisioned a scheme in which the forest industry would be nationalized, then rationalized, so that the cut would be controlled, the government would get the profits, and the cut could be reduced as needed. Sloan had brought in a plan that did increase government control, but it also helped the companies then operating. They were the firms that had "first cut" at the available wood and that could apply for Tree Farm Licences. When operating on PSYUs for several years, they could shrug their shoulders and say that reforestation was the government's business, not theirs.

Whether Sloan was "just being realistic" or "the best friend big business ever had" is now unimportant. The same applies to whether Cameron was right or whether he was a beer parlor idealist. Sloan's (or Mulholland's) plan officially initiated sustained yield forestry in British Columbia, without more than a ritual whimper from industry. Although future governments could modify Sloan's ideas, they couldn't erase them, since they quickly became the basis for investment, jobs, and a bureaucratic structure within the Forest Service.

It must have been a painful lesson for Cameron. His suggestion of a "compulsive and integrated state scheme" demonstrated something known to merchants since the invention of war: nothing provides the opportunity for reaping profits at public expense better than a noble cause.

Chapter 10
The Postwar Boom

Spring break-up at Newlands. Photo by Harold Mann, Courtesy of Prince George Public Library.

When Ray Williston arrived in Prince George toward the end of 1945 to become the first Supervising Principal of Schools:

> ...there were no rooms available. Finally, the owner of the Prince George Hotel took pity and gave me an army cot set up in the hall...for 75 cents.
> At the time I arrived, the army was disbanding the Sixth Canadian Division and everyone thought that when the army left, Prince George would fold.

It didn't. Instead mills started up, then kept going, then more mills arrived. Apparently every Canadian and every American soldier in Europe spent his evening discussing tactics of the timber trade rather than those of the military. The 1947 annual management report said:

> The influx of new sawmills continued unabated during the year with a total of 104 new mills established.... The mills along the C.N.R. increased to 258.... During 1947 mills in the vicinity of Quesnel increased to 61, almost double that of 1946, and there is every indication that more mills will be starting up.

But most of these mills were very small. Many were farm mills that operated only a few months of the year. Others were portable mills used to salvage timber after forest fires. The 382 mills listed in the report included 53 that weren't operating. The number of board feet cut ranged from 6,000 at H.V. and J.V. Davis's mills to 11,389,000 at Eagle Lake.

Mostly the mills manufactured junk. The China market wouldn't take it. In 1948 the South African market refused to accept such sloppily manufactured lumber. Until the end of the 1950s, there was good reason lumber from the area was called Northern Inferior.

In 1945 the population of the area was less than 4,000, including the lumbermen who ran these mills, twenty-seven teachers, the staff at two banks, five car dealers, the staff of eight licenced hotels, the employees of five wholesale distributors, and forty-four people on staff at the Forest Service for the district. When you consider the number of wives and children, it is obvious that the average sawmill employed fewer than a dozen men.

To supply the mills, the Forest Service put up blocks of timber in the PSYUs for auction. In 1946 the average sale worked out to just over 219 acres per listed mill, for a total of 53,049 acres, and the price received for cutting rights averaged $11.20 per acre. The next year just over 58,000 acres went to auction, and the price rose to an average of almost $17 per acre. In 1948 the price jumped to over $20 per acre for 60,770 acres. This meant that for the first time the Forest Service received over $1 million in revenue from sales in the Prince George District.

What were the mills doing to keep up with the times and these changing prices? Western Plywood at Quesnel was using a converted army half-track to yard logs out of the bush. Trapping Lake Sawmill at Red Rock was using a self-loading trailer mounted on four war surplus army tires. A lot of tanks were bought, their turrets were removed, and they were used instead of Caterpillar tractors—the tanks were cheaper. Carrier Lumber has a tank transporter that was used to haul logging equipment. You can still get parts for a Sherman tank from a Vancouver logging equipment supplier.

Heavy equipment was in short supply, so people improvised. World War II had taught the forest industry the advantage of mechanized equipment. As a result, the number of tracked vehicles increased, while the number of horses went down. Logging trucks grew in number and size. Arch skidding—in which a log is cranked up against a truck so that one end is in the air and the other end drags on the ground, like a car that's being towed away—made its first appearance in the north in 1947.

Ernie Anderson skidding at Giscome. Photo by Harold Mann, courtesy of Prince George Public Library.

The rough lumber mills that handled the logs had changed little from the 1930s. But at the larger mills a new building was added to house the planer, which planed the rough lumber smooth to precise dimensions. Some independent planers were built to handle the output of twenty or more rough lumber mills. There had been fewer than half a dozen planers in the north before the war, but in just a handful of years a cluster sprouted in Quesnel and seventeen on "Planer Row" (River Road) in Prince George. By 1954 the number grew to almost a hundred, which processed seventy-five percent of the lumber produced in the Prince George District.

Some of the new sawmills and planers were offshoots of coastal mills that expanded into the north. Others were started by newcomers to British Columbia from Europe. Still others were started by prairie operators who had pulled up stakes and moved west.

Roy Spurr sold out at Eagle Lake, though he stayed on as manager, making a million dollars in the process. He invested it in another mill, of course.

To handle the almost 400 mills and a forest area of 23,097,800 acres—bigger than Ireland, Austria, the United Arab Emirates, or England—the Forest Service employed a full-time staff of fewer than two dozen people, plus seasonal rangers, patrolmen, and lookouts. Though the staff worked frantically, it couldn't help but fall behind.

Since the early days of logging in Michigan, a common strategy had been to get the rights to some land at the mouth of a valley, then pull all the timber in the valley through it. Even though they were overworked and making several inspections a day, the rangers were able to catch a lot of this overcutting. They fined one McBride area operator for cutting an extra 227,594 feet, board measure, of sawlogs, caught the Pouce Coupe operator who cut a quarter million feet more than he'd bought, and found thirty-nine other trespass cases that added up to almost a million and a half board feet of sawlogs, plus ties, pit props, and firewood in a single year.

For the year 1948, the Forest Service reported that the number of mills along the CNR had declined to 234. "This was due in part to the depleted

Postwar innovation: Western Plywood's converted half track used for skidding logs and loading trucks near Quesnel in 1947. Photo courtesy of B.C. Forest Service.

supply of roadside timber, poor weather conditions, increased cost, and falling markets later in the year." In addition, like every boom, from a gold rush to the development of integrated circuits, logging attracted a lot of people who should have stayed home. The annual management report went on to say:

> *The bulk of the orders to the U.S. call for dressed lumber which must be dry and the industry is not geared to produce dry lumber. There are only four companies operating kilns in the district.... Because 1948 was a poor drying year with a wet summer from July 1st on, many operators could not pile dry successfully.*

Today most operations in the north have kilns, and now the lumbermen are finding they have to turn them off when cutting boards for export, since kiln-dried wood may mean a higher tariff in Europe. However, in the postwar period it was common to see lumber mill yards that resembled scale model metropoli; their carefully piled stacks of boards looked like modern skyscrapers, each twenty or more feet tall.

In 1949 the weather was wet enough to make some people consider raising mildew commercially. Lumber prices were down. As a result, the number of acres involved in timber sales went from over 61,000 to less than 49,000. The average price bid dropped by fifteen percent. Just one percent of the wood cut was exported overseas, while forty-five percent went to eastern Canada. Twenty-seven mills didn't operate during the year, and many of the others cut for only a short time. A lot of mills changed hands. Fifty-four were deleted from the mill list for various reasons: several were dismantled and set up in other parts of the province; some just went under. Every week the newspapers throughout the north carried ads of mills for sale.

That was the first year since the war that the mills had lost money. Prices for lumber have gone up and down ever since, and analysts have plotted the number of years between cycles, trying to forecast the future. Conventional wisdom today says that a bad market lasts at least two years.

However, in 1950 prices reached a record high. Timber sales covered 12,000 more acres than in 1949, and the average price paid for timber climbed $198 to $3,628 per sale. The number of mills went up too. By year end, over 500 were registered. Provincial forests were staked, and Public Sustained Yield Units began to become a reality. The boards, timber, ties, and other forest products produced were worth over $1.25 million in direct fees to the provincial government. In 1951 provincial revenues went up again by forty-nine percent. Timber sale prices rose by fifty-eight percent, the average price paid jumping to $7,059. The forest products produced in the Prince George District were worth over $34 million.

Although the industry was booming, it was still an erratic one. Of the 3,185 men employed in the industry in the north, a third worked between nine and twelve months that year, and 1,324 had jobs less than half a year.

The millowners who made money, and who had foresight, invested their profits in timber. They had to. The government was setting up provincial forests that contained Public Sustained Yield Units, following Sloan's suggestion, and Tree Farm Licences. As a result, the total amount that could be cut per year was restricted. Because limits to growth were in sight, buying timber at the time ensured that you would be able to stay in business down the road—and determined how big you'd be when all the timber had been allocated. The operators complained about this, of course, but rushed into the first PSYUs set up: Crooked River, Naver, and Cottonwood, plus Stuart Lake. The first Tree Farm Licence, given to Weldwood in November 1950, helped the operators "see the value of sustained yield forestry," in the words of the annual management report.

That's a euphemism. In administering the PSYUs, the Forest Service tried several never-exactly-defined approaches. What they boiled down to was: if you don't have cutting rights, you can't get cutting rights. If you have cutting rights, you can get a replacement as they run out. The amount you can get depends on what you already have. The Forest Service used phrases like "established operator" and "committed annual allowable cut" or "licence priority," but what they really meant was a timber quota, with each established operator getting a percentage of the allowable annual cut based on what he had cut during the previous few years. If one firm bought another, the quota was transferred to the firm that bought. As you'll see later on, this quota system led to a thirty-year argument that still continues.

Then in 1952 the arrival of the Pacific and Great Eastern changed the situation, ending the postwar boom, and replacing it with a new era: the era of big money.

Loading at Sinclair Spruce Mills. Photo courtesy of B.C. Provincial Archives.

Chapter 11
The Pacific and Great Eastern Arrives—Forty Years Behind Schedule

> *I am not going to become foster father to this illegitimate offspring of two unnatural parents. It was a waif left on my doorstep. It was conceived in the sin of political necessity; it was begotten in the iniquity of a half-million dollar campaign fund. I refuse to be the godfather of any such foundling.*
> — Premier "Honest John" Oliver, 1918

Most people who have looked at the history of the Pacific and Great Eastern, now the British Columbia Railroad, haven't been any kinder than Premier Oliver. The most recent of several Royal Commissions that have looked at the BCR said:

> *At the risk of straining Premier Oliver's metaphor further, we must observe that we have difficulty in distinguishing between the times of conception, gestation and maturation of his "foundling". We know that 44 years were to pass, between 1912 and 1956, before a train could run all the way from North Vancouver to Prince George, still only a partial realization of Sir Richard McBride's total scheme for a railway from the 49th parallel to the Peace River District and ultimately the Yukon and Alaska. Despite his denial of paternity, Premier Oliver's government saw no alternative but to maintain the child....*

From the passage of the original bill in 1912 calling for the railway to link the Grand Trunk Pacific (now the Canadian National) in the north with Vancouver to the election of 1915, progress looked good. The promoters, Foley, Welch, and Stewart, had helped to build both the Grand Trunk Pacific and the Canadian Northern Railway. The new railway's debenture issue sold out in London. By 1915, 176 miles of track had been laid from Squamish to Chasm. But the finances of the railroad were a mess. A government inquiry in 1917 determined that much of the fault was due to the previous Conservative government—perhaps because the three Liberals involved in the inquiry could outvote the two Conservatives. Despite his protestations, Premier John Oliver took over the railway in 1918. In 1919 track reached Williams Lake. In 1921 it reached Quesnel. Then it stopped for almost thirty years. Finally, in 1952, the railway reached Prince George and stretched south to North

Vancouver four years later. Immediately there were demands to extend it. The government expanded it northeast to the Peace River block and past Fort St. James toward Driftwood Valley to the northwest. By now the role of the railway had changed. As the 1978 report said:

> *The railway was seen as a "development tool". This is not a precision tool, made to fit a special bolt, but an all-purpose instrument thought by optimists to be capable of opening everything that is closed, including minerals locked into the ground and remote forests far removed from economic exploitation....*

After noting that eighty percent of all carloadings were derived from the forest industry, though the industry paid less than sixty percent of the revenue from freight, the Commission asked the Council of Forest Industries if it would take over the railway since the government was forced to—in effect—subsidize the industry.

In several thousand well-chosen words, the Council said, "No way."

What has the railway to do with the forest industry in the north? Everything.

The Canadian National's "East Line" from Prince George to the prairie provinces let the industry get started in the north. As the "Prince George Eventually" inched its way north, mills were built along the right-of-way. Today the only unexploited forests in the province are where the railroads don't go. If the forest industry had to pay what it costs the BCR to haul lumber, the mills in the Interior would be—from the industry's point of view—"unable to compete in the U.S. market." However, this argument ignores such variables as the changing value of the Canadian dollar.

In 1972 there were forty-seven sawmills of major size along the BCR. They produced a total of 3.025 billion board feet of lumber, twenty percent of the total provincial lumber capacity. The nine Interior pulp mills constructed since 1962 accounted for a quarter of the province's total pulp output in 1972, and many of these were on the BCR. Back to the Royal Commission on the BCR again:

> *The provincial government had, and still has, the power to influence the location of forest industry development.... (It) owns practically all forest land, decided where and when resources would be made available, demanded sawmilling and pulping facilities should be joined to every timber right issued.... At Prince George, a community served by the Canadian National Railroad for 40 years, before the British Columbian Railroad arrived, the new sawmilling and pulping facilities were constructed on trackage of the provincial system....*

Even before the railroad arrived in the north, it had an effect on the area. P.T. Barnum once said, "No man ever got poor underestimating the intelligence of the American people." Judging by the number of times the railroad was announced, and the people reacted, the same is true of British Columbians.

The full story of the bungling and cupidity and the starts and stops of the Pacific and Great Eastern would make a good television comedy. In 1897 the Cassiar Central Railway was incorporated to build a line from the Stikine to Dease Lake. Three years later the Vancouver, Westminster, Northern and Yukon Railway was incorporated to build and operate between Vancouver and Dawson. In July 1913 Premier Richard McBride announced that the Pacific and Great Eastern would run from Panama to Alaska. In 1942 a railway from Prince George to Alaska, via the Rocky Mountain Trench, was surveyed by the U.S. Army Corps of Engineers. The Alaska highway was chosen as a quicker-to-build, less expensive alternative. In 1943 the Postwar Rehabilitation Council of the B.C. legislature, headed by Prince George MLA Harry Perry, proposed that the PGE should go to the Peace River district, then to the Yukon and Alaska, as well as south through Vancouver and into the United States.

In 1948 the U.S. House of Representatives Foreign Affairs Committee voted to negotiate with Canada for a rail link to Alaska. The Spokane, Washington, *Spokesman-Review* announced on July 29 "...that the American group is negotiating with the province of British Columbia for purchase of the Pacific and Great Eastern." The Canadian government said it wasn't interested. On August 12, the Prince George *Citizen* reported, "President Truman stated today that he considers the Alaska railroad very necessary and he will initiate discussions with the United States defence minister...." Ignore the fact that in the United States the only ministers are in churches and the title of the official in question is Secretary of the Department of Defense. The main point—and a main point in the development of the entire pulp economy of the north, as we'll see later—is that the United States wanted the railroad for military use.

On October 14, 1948, the Prince George *Citizen* carried the headline:

U.S.A. READY TO START ALASKA RAILWAY
Private Capital Available to Build
Greater Part of the Project

However, Willis T. Batcheller, who headed the group that wanted to buy the railroad, wrote to the *Citizen*, "I am informed by Ottawa three years ago the decision was reached that neither the Dominion government nor the Canadian railroads were interested or in a financial position to extend the P.G.E. railroad." Early in 1949 Premier Byron Johnson announced, "I have received no bona fide offer for purchase of the PGE and the government is not prepared to alienate the resources of the province for any promotional scheme." Another one bites the dust.

On June 2, 1949, sealed tenders for construction of the last lap to Prince George were opened. In 1952 the railroad reached the city.

Then in November 1956 Premier W.A.C. Bennett and Ray Williston signed a letter of agreement with Axel Wenner-Gren to allow investigation of resources in the north. Part of the Swedish financier's proposal was a 180-mile-per-hour monorail from Prince George to the Yukon.

On May 24, 1960, the Pacific Northern Railway was incorporated. Ten men and three pieces of equipment were in view a month later when the Premier, official party, and a gaggle of reporters arrived by plane at Summit Lake. A tree was cut with a power saw, some dirt was removed, and then the group enjoyed prime rib, potato salad, diced watermelon, and salmon. After lunch everyone—including the work crew—went home. In the fall the Public Utilities Commission dismissed the application.

In the late 1960s the railway was pushed to Fort St. John, then north to Fort Nelson. In 1971 Premier Bennett, some U.S. railroad men, and various politicians made a special trip to inaugurate the new extension, which, in the swamps, had been laid on gravel that rested on plywood. The government hadn't provided enough money to build the route properly. But even before the completion to Fort Nelson, Premier Bennett had another idea—to extend the railway northwest to Alaska. He chose the route during an afternoon helicopter ride, and in 1969 construction crews went to work.

The election of the NDP in 1972 didn't halt construction: Moira Farrow of the *Vancouver Sun* and some biologists who risked their jobs did. After an extended series of articles revealed the environmental devastation threatened by the railroad's newest extension, the NDP called a temporary halt to construction. Then word came out about poor construction practices. When the NDP government was defeated by Social Credit in 1975, the extension—allegedly going only to to Dease Lake—came to a halt. After the 1977 Commission report, it was shut down.

The story isn't over. Every British Columbian can expect to see the railroad to Alaska announced again...and again...and again.

Nonetheless, the arrival of the Pacific and Great Eastern at Prince George, November 1, 1952, meant a faster, cheaper way to ship lumber, with new points of rail access for mills. And the boom changed character again.

Chapter 12
The Messiest Strike in Western Canada

In 1952, B.C. lumber was gaining acceptance in the central United States, creating new markets for northern mills. Seaboard Lumber Sales, the cooperative exporting firm founded in 1919, was selling Interior lumber overseas. Six hundred and four mills operated, but another ninety-six sat idle because, while the market was good, it wasn't as good as in 1951. The Prince George District had eighty-nine planers, most owned by companies that also produced rough lumber. At year end, it looked as though 1953 would be another year of expansion and complaints.

In the northern part of the district, companies were using crawler tractors to skid logs to landings where they were loaded onto trucks and taken to the mills. Near Quesnel rubber-tired skidders were used. The Peace River area and the light stands near Vanderhoof were the only places where horses were still popular, according to the annual management report: "The scarcity of competent teamsters limits the use of horses. Now the one-man power saw is the standard falling and bucking tool." The biggest complaint in the report was that logs were shipped from Quesnel to the Powell River Company, then chipped for pulp. The Forest Service said these "operations are now carried out to cut immature trees and (they) create excessive slash." There were two million acres in a dozen Tree Farm Licences in the province, and Premier Johnson rejected the suggestion that they should be managed in detail by the Forest Service. Instead, he decided that the industry should do the work and the Forest Service should regulate the industry.

In 1953 more mills went into the area: 18 in the Peace River district, 13 in the Quesnel area, and the East line broke 400, to bring the total to just over 700. The area logged went up to 55,724 acres.

The market was good during the opening of the year, then softened later on. This softening pleased some millowners, who believed that when it came time to negotiate a new contract with the IWA later in the year, the poor market would help demonstrate management's inability to pay higher wages. The millowners, who had formed a branch of the Canadian Manufacturer's Association called the Northern Interior Lumbermen's Association (NILA), hired Bob Gallagher as the Association's first full-time manager.

It seemed that a strike could be easily avoided. Only twenty percent of the 3,155 men employed in the mills and the woods were IWA members, although the union did represent the workers at most of the big mills in the district. The union realized that when the market was poor, management could afford to close down the mills rather than run at little or no profit. Mill managers calculated that if a strike did take place, it would be of limited duration because neither the union nor the workers could afford to continue through the long winter. Then, when strike funds were depleted, management would be able to get what it considered a good settlement. In addition, a forty-two-day strike on the Coast had scared union members in the the north during the previous round of contract talks, so management felt it had a good chance of talking about the effect of a strike on the individual worker and avoiding one this time.

The union asked for 18 cents an hour, more paid holidays, and union shop. Ever since the IWA had started enrolling members in the north, management had refused to "check off" union dues and thus did not automatically collect them as a deduction shown on paychecks. That meant the union had to collect the money from the workers each month. (Surprisingly, only one month's dues, from a single mill, were paid to a tavern rather than to the union office.) And although the majority of men at a mill might join the union, and the IWA might become the official bargaining agent for the workers, there wasn't any rule in the contract stating that a worker had to join the union in order to work at the mill. The union felt that this gave men who opted out a free ride, getting the benefits of union membership without having to pay any of the costs involved.

The Northern Interior Lumbermen's Association, which represented the biggest operations in the north, responded with an offer to extend the old contract, without changes.

The conciliation board suggested a six percent increase to raise the base pay rate from $1.30 to $1.38. The union accepted this offer. Management rejected it.

When negotiations begin to get difficult, it's usual for a union to call for a strike vote. This doesn't mean a strike. Instead it lets the union negotiating team demonstrate that the members are behind it and are willing to strike if necessary. In 1953 the union had to vote to hold a strike vote and then hold a government-supervised strike vote. Both passed.

As soon as the strike vote results came in, the men at Prince George Planing Mills walked off the job. The night shift went down to the union office and told Mike Sekora, the union organizer, what they'd done, then passed the word to workers in other mills. Sekora called an immediate union meeting. The same day the Prince George *Citizen* announced:

IWA STRIKE ORDER
IDLES THOUSANDS

No Break Seen in Deadlock;
Pickets Patrol Mill Areas

> *Prince George's "planing mill row" was ominously silent this morning as a lightning fast strike of the International Woodworkers of America workers descended on the Northern Interior lumber industry.*
> *Not a wheel was turning and the rain-drenched pickets huddled in doorways beside rapidly deteriorating cardboard signs which read, "This plant is on strike."*
> *...Throughout the city of Prince George, a gloomy outlook prevails among merchants and businessmen as they watch the first signs of deterioration of the Interior's lumber industry....*

Neither side would give an inch. Each side was sure it was "right." Each side had feisty members who preferred to fight rather than to talk. Most of the millowners were ex-woodworkers, and there was no way they were going to give up what they'd fought so hard to get. They rightfully saw most of the workers as men who drifted in and drifted out, following their whims, but they ignored the fact that the lumber industry had always been this way. The men looked at their bosses and said, "They're no better than we are. Hell, just a few years ago they were what we are. Why should they get so much profit?"

The union members were angry—and confused. It's normal when a strike starts. You wonder about money. You wonder what's really going on. How long will the strike last? What's the strike pay? What am I supposed to do? You go down to the union office and everyone's there. Lame jokes. People moving fast through the crowd—are they really hassled or are they just trying to look important? Some guys have instant speeches. Some receive a nod and go out in the alley to talk to union officials. It's the start of a strike, but you feel as though you've walked into a play halfway through the third act. Finally you're given a nod. You're to picket for eight hours on, then get twenty-four hours off. This means it'll be hard to get a picture of the whole strike. There's a meeting tomorrow night, 7:30 at the CCF Hall on 4th Street.

But that still leaves too much time. Time to sit and worry about car payments, house payments, how you can cut back, whether you should stock up on food or try to get a job elsewhere.

That night, September 28, a lot of husbands and wives stayed up late, talking and worrying over something they couldn't control. Some people stocked up on canned goods the next day. Some bought cigarettes. Some stocked up on booze that didn't last as long as they thought it would.

But the lights burned late at millowners' homes too. The strike was unexpected. Negotiations really had just begun. When the mills could no longer ship, what would happen to the customers that had been so carefully cultivated? The customers would seek alternate sources. Would the union really be able to shut down the nonunion mills? If not, perhaps they could ship a bit to keep the customer list active. For many mills with a lot of payments it's better to keep running, even at a slight loss, if it means you can pay most of the bills coming in rather than none. By the second day of the strike, millowners were talking to bankers and to their accountants, trying to find ways to survive.

By the third day of the strike, the men on the picket line could eat at a soup kitchen set up in the CCF hall that served moose stew three times a day. The moose supply lasted throughout the strike. After the hunting season was over, the phone would ring late at night at the hall. "There's a moose out back. Do you think you could get it in off the street?" The local conservation officer made midnight deliveries of moose that had been hit by trains or cars. Farmers offered free vegetables if the strikers would pick them. A ton of potatoes was collected. Within a week the number of men reporting for picket duty had grown from 200 to over 400, including workers at nonunion mills who had been thrown out of work by the strike. The only way to be allowed to eat at the soup kitchen was to picket, so they picketed.

Under the labor legislation of the time, only unionized operations that had voted in favor of strike action could be picketed. It was clearly illegal to close down nonunion mills, or unionized ones that had voted against strike action. But since less than twenty percent of the workers in the industry had actually voted in favor of the strike, following the rules would mean losing the strike. So the union sent its pickets to every mill in the area, to the rail yards, to every place that might have shipped a board. And workers outside the union, looking at the lines of pickets, realized they too were out of work. The north shut down.

For the first month it was largely a paper war, with charges and countercharges appearing in the newspaper and the Northern Interior Lumbermen's Association refusing to go to meetings. The Bay's sale of pullovers for $4.99 was ignored. No one on either side had money to spend, since no one knew how long the strike would last and coffee was up to 93 cents a pound.

Mass meetings run by the IWA to show support for the strike regularly drew 500 people. The *Citizen* commented:

THE LIGHTS ARE GOING OUT

> ...Already a great deal of bitterness has been engendered between the rival factions. The operators charge that the union leaders manipulated the strike vote in a manner that made it possible for a minority of members to tie up the entire industry and claimed that "Communist sympathizers" dominated the IWA strategy board.
>
> Union members retaliated by asserting that well-heeled members of NILA welcomed the strike as a means of forcing hundreds of small operators to the wall and as a chance to "wreck the union"....

The stakes were high. If the IWA won, it would add many more operations, since nonunion workers would see what a good job the union had done for its members. NILA was as weak as the IWA, representing just thirty-five of the eighty-five biggest operations that had been shut down; if it won the strike, it would end up with a lot more members.

It made sense for NILA to freeze out the IWA.

It made sense for the IWA to use intimidation to win the strike.

Of course it would have been bad taste to make these things public, so the strike operated on several levels at once.

The public version was set out for worried merchants to read in the newspaper: advertising by NILA, letters to the editor by both sides, and news stories like "Mills Would Lose Money at Today's Prices: NILA" or "Small Independents in Fear of NILA Reprisals," which quoted Fred Fieber, who had taken over running the strike for the IWA. Alex Bowie, head of a special committee of the Board of Trade, tried to get both sides of the dispute together at a meeting. NILA refused to come and wouldn't go to a meeting the Labor Relations Board mediator called.

The private version of the strike was $8 a week in strike pay for single men and $16 per married couple, with a little extra per child. To be able to pay that much, the IWA encouraged single men to leave the north and get jobs elsewhere. The union rented shacks for men from outlying mills that had closed down. Rooms were subsidized at hotels, with four men to the room. Roy Yip of Royal Produce saved hundreds of families from starving by giving credit to IWA members. Car dealers threatened to repossess until Howard Webb said to the local dealers' association, "You want 500 cars back on the lot tomorrow?" The dealers agreed to suspend payments for the duration of the strike. New ads began to appear in the newspaper offering "the new 1954 models with no payments until February." Finance companies got calls saying, "Come over and repossess my stuff. I'm leaving town—for good."

Small sawmills closed down without having pickets appear. One reason was that the mills felt threatened by NILA; if they operated, they'd be paying workers, and that might mean money that could prolong the strike. But they also felt threatened by the IWA; if they shipped limber, it could be used to "prove" the industry hadn't really been shut down. So they closed up and waited.

Larger sawmills either closed down or tried to keep going despite the strike. The ones that tried to run were quickly closed down by groups of men with two-by-fours who suggested it would be healthier to stop work.

Every other local in the IWA chipped in money. One of the members of 1-424, Carl Mueller, took a truck down to the Coast and toured logging camps. He came back with almost two tons of clothing, mostly for the children.

Strike pay came out on Monday morning, and by noon some of the men were broke and drunk. Harry Neal, who was on the strike committee, said, "On Monday morning incidents were likely to start. The guys would get a few drinks and get nasty. So I never went into a bar during the entire strike."

By the end of the month, most woodworkers, and some of the millowners, were flat broke. The strike had been so unexpected that word didn't reach some operations on the East Line until the strike was a week old. No one had been able to save in anticipation of the strike. A lot of people had to leave the area just to survive.

Even today people are reluctant to discuss the third level of the strike: the war. "You don't want to talk about the time you came home and found everything in the house ripped apart and shit smeared over the walls, you don't even want to think about it," a management member said.

Management knew the IWA couldn't survive a long winter's strike, so its strategy was to sit back, refuse to talk, and watch men leave town. Eventually, the managers figured, the union would run out of cash, the men would run out of cash, and they'd get the previous contract renewed without any changes. With the union financially exhausted, there would be little possibility of a strike for the next several contracts to come.

The union strategy was equally obvious. It had to shut down the entire north and keep it shut until pain in the pocketbook forced management to give in.

The IWA sent Jake Holst to Prince George to sort out the strike. He said:

> I was walking down Third Avenue with one of the men in charge of the strike. He saw a member of NILA across the street, stepped off the curb and yelled some sweet things at him. I said, "If you want to say that in private, that's fine, but for heaven's sake, don't yell it out on the street." He replied, "That's how I feel and that's what I'll say."

The conflict among strike committee members in the union wasn't over what should be done; it was a conflict in style and degree. Some members were broke, frustrated, angry, and out for blood. Others knew that the fastest way to get public opinion against them, and lose the strike, was to appear brutal. So after a rash of incidents, the strike committee was reshaped, then reshaped again. At various times Mike Sekora, Fred Fieber, and then Jake Holst headed the IWA's effort.

None were successful in containing the violence that breeds in hunger and in bottles.

Goon squads went against goon squads. If only two or three men were on the picket line, they would get beaten up, so a dozen had to go together. Mill management had to travel in packs. At times it wasn't safe to be the wife or child of a member of management. Homes were vandalized. Within a week after the strike had started, management had two "security groups" in town. A week later the leader of one of these groups was eating at the Club Café when John Hemming, a union member from Sinclair Mills, came in.

"Sullivan," John said to the leader, "clear out before we come after you!"

Sullivan looked up from his coffee. He had ten men in the café. John was alone. Then Sullivan looked out the window. John had told some of his friends what he was going to do. Though John didn't know it, his friends packed the sidewalk. Sullivan and his friends left town.

Management responded by bringing in another squad. To this day some of the loggers they beat up argue about whether they were Pinkertons or ex-RCMP officers. Whoever they were, they were professionals who knew how to beat a man until he passed out without leaving a mark on him.

Union tactics were no better than management's. One millowner's wife and children were terrified by a phone call from a unionist. The next day the millowner saw the unionist on the street. According to five people interviewed, he felled the man with a single blow, then picked him up and kicked him, then picked him up and kicked him—for three blocks down Victoria in Prince George.

"A bunch of us were picketing at a mill when a management man drove up," one logger recalled. "We stopped his car, then started rocking it. We would have flipped it over—we had it rocking so it was way up in the air leaning on two wheels—when the RCMP showed up."

Why did you do it?

"Oh, you know how it is when you're young, single, and there's a bit of devilry that can be done. We were all young and foolish."

"I lived out on the East Line back then," another man said. "The strike caused hard feelings that are still there. It took years before people would speak to each other. Management people came home and found strikers' wives—they were the worst ones—writing things on the walls in their children's shit."

Trying to sort out the stories even after almost three decades is almost impossible. One millowner, alone in his office, saw cars full of picketers driving to the mill. His story and the union version differ wildly. Only three things were the same in all versions of the story: the strikers "treed" the millowner in his office, he ended up firing over their heads with a rifle, and in 1980 there were still hard feelings on both sides.

In the beginning, the "official" groups of union militants limited their activities to felling trees across roads leading to mills and standing around with two-by-fours in their hands, thus discouraging nonunion operations from running.

Then things got rough.

The employers went to court every time the IWA violated the rules. In all, there were twenty-seven injunctions issued against the union. For its part, the IWA figured that since it was outside the law, it couldn't go to the law. Besides, the local had only one lawyer, Alex McDonald, who later became Attorney General of the province. Alex was busy freezing in a tar paper shack in Williams Lake while discussing the law with one group of men, then freezing in Hutton as he went through it again with another group. "The real issue wasn't union shop or wages," he said. "It was union recognition in the north."

And the union had no alternative to intimidation.

"The fight," Jake Holst said, "was to make management realize they were going to have to live with the fact of the union, like it or not."

Management didn't like it. Antistrike demonstrators clashed with union pickets in Quesnel. Both sides got rougher. Neither side had "right" on its side—or wrong. Neither side acted responsibly.

Premier Bennett tried to settle the strike. He had meetings with both sides, but NILA rejected his plan before it was even put on paper.

The RCMP was walking a tightrope. Members who had been in the north and who knew the men looked the other way. Detachments shipped into the area—"the Cossacks," as the strikers called them—felt the strikers were revolutionaries. Those in charge of the Mounties tried to defuse the situation, working with both sides to avoid problems.

But there were problems. And they began making the newspaper in late October. At the end of the month, strike committee member Howard Webb told the *Citizen*, "The union does not condone the initiation of violence, but we are ready to resist it where it is encountered. If anybody gets hurt, we are going to see it is not us." And since Howard was one of the more responsible union members, respected by the community, his statement was taken very seriously.

On November 6, the Canadian National got an injunction to stop union pickets from halting rail shipments. Prince George Planing Mills and Penny Spruce Mills received injunctions to halt all picketing. Rustad Brothers—who had obtained union permission to rebuild their mill—got an injunction to allow lumber production without hassle. With mills starting up, more violence started as well. Twice, when people working at the mills were beaten up, the incidents hit the papers. Because the union was still avoiding publicity when their people were on the receiving end, the stories in the newspaper suggested that only the IWA was acting unreasonably.

Meanwhile, the population of Prince George was dropping. A few weeks after the start of the strike, merchants reported sales down eighty percent at a gift shop and twenty-five percent at a hardware store; even grocery sales dropped fifteen to forty percent. Businesses that were marginal went under. Store owners cut staff, then cut wages for the remaining employees. By the end of October, the newspaper stopped printing stories about businesses going broke. No one wanted to let the rest of the world know that the city was shrinking. By the beginning of November, almost a quarter of the town's

population had departed, including those who had watched their retail businesses die because of lack of customers. Alex Bowie, A.B. Moffat, Eugene Habriele, Bill Munro, John McInnis, and others on the Board of Trade's special committee, plus Ray Williston, the MLA for the area, tried to bring both sides together. This exercise in stubbornness was killing their town. Small mills were going broke. Sheriff's sales were common. And Prince George was getting a reputation as a nice place to avoid visiting.

According to Tage Mogensen, who later became financial secretary of the IWA in Prince George:

> *People were frantic. They couldn't feed their families, they couldn't get jobs, they had no hope. For the first time we began to have trouble with people stealing from the soup kitchen. If your children are really starving you do things you'd normally be shocked by. I had money in the bank and planned to buy a new truck for Christmas. Then the strike started. When it was over I had less than $100 in the bank and no truck.*

On November 23, the Princess Theatre showed *The Titanic,* starring Clifton Webb, but no one had money to see it.

By the end of November, those who were going to leave town had left. Both management and the union knew that the men left would keep picketing as long as necessary. There was a rumor that the strike was almost over. But it proved to be only rumor. The IWA, the Canadian Legion, and several churches and service clubs got together to "make sure that food, clothing and toys will be plentiful in every Prince George home throughout the festive season," as the *Citizen* put it. Arnold Webster, head of the CCF and Leader of

Photo courtesy IWA Local 1-424.

the Opposition, came up, looked the situation over, spoke in favor of union shop, and got out.

On December 3, negotiations resumed. On December 7, the union rejected NILA's offer of 5 cents an hour—effective the following April. Instead of holding check-off for the entire area in a union shop, NILA was willing to do it plant by plant. That meant that if a man changed jobs, the union had to sign him up all over again.

Judge Arthur A. Lord agreed to act as a one-man industrial inquiry commission and came north on December 14. NILA ran a full-page ad in the newspaper saying there was still time for a payday before Christmas if the union members acted fast. Judge Lord gave both sides hope that the strike would be resolved. A little lumber began to be shipped by companies protected by injunctions. And some union members persisted in making threatening phone calls to nonunion workers and railroaders, even though NILA publicized every one of them.

The town was running out of steam—and money. A jewelry store had a twenty-five-percent-off sale the week before Christmas to try to stay alive. Employees at a car dealership volunteered to have their pay cut by a third so the staff wouldn't have to be cut.

On December 21 the Lord Report was released. Although it offered 5½ cents an hour effective immediately and a watered down union security clause, one part read:

> *...To my mind it is impossible to suggest the union is a responsible union and is entitled to compulsory check-off, or union shop. Affidavits list such serious unlawful acts, including violence, assault and damage to property on the part of members of the union as to show a complete disregard for the laws of this country.... They should realize that flouting the law is only a detriment to their cause.... The union must realize they can not attain their ends by unlawful acts.*

NILA accepted the report. The union rejected it. As "Perambulant Penman," a *Citizen* columnist, put it, "The Lord Report said the right thing the wrong way. You don't insult a customer you're trying to sell something to." The Lord Report helped turn public sentiment against the union. The day it came out, all hope of a union victory was killed by a man who had been fired by Northern Planers and three of his friends. Underneath a huge "Merry Christmas" banner, the *Citizen* reported on Christmas Eve:

BRUTAL ATTACK SENDS WORKER TO HOSPITAL

> *Lying in a Prince George and District Hospital Ward bed today, his face beaten to a puffy mass, his nose broken and possibly his jaw, is a slightly built 26 year old Dutch immigrant who Monday night was the victim of four hoodlums alleged to be members of the union violence squad.*
>
> *Through swollen lips and aching face Fred Tilstra, a planing mill employee, told a **Citizen** reporter Tuesday how he was viciously attacked by four men on the outskirts of the city around 7 p.m. and beaten into unconsciousness before being left bleeding beside the road....*
>
> *Throughout the half hour interview he anxiously*

scanned every visitor who came through the door of the ward and twice while talking to the reporter he said he is afraid, "the union will send somebody...."

This wasn't a planned union move. But it had happened. And the first person arrested for questioning was a union member.

The afternoon the story broke, there was a Christmas party for the strikers' children. In photos of the party the children look happy. Their parents look grim. The adults at the party knew the strike had been lost and there was nothing left but speeches and the signing of a contract.

A few days later a citizens' meeting asked the provincial government to enforce the law in the area. Outgoing mayor Garvin Dezell said, "The riot act will be invoked unless order is preserved."

On New Year's Eve, the union met with NILA until ten at night to work out a contract, aided by Ray Williston and members of the Board of Trade. On January 5 the contract—almost exactly what the Lord Report had recommended—was signed.

All that was left was for businesses to continue to go bankrupt for the next year as a result of the crippling effect of the strike, for people to gradually get to the point where they could simulate friendliness (though some on each side can't do it to this day), and for both the union and the millowners to try to get the industry back on its feet again.

Chapter 13
Recovery and Prelude: 1954-1956

The strike shook the confidence of Prince George, but not the confidence of the north. In August 1954, *Truck Logger* magazine reported:

> ...It's the same as '98 was to the Klondike!
> There's a rush on, but it's divided into organized moves—unlike the helter skelter of prospectors after a quick dollar!...
> **Levelling Off**—Prince George, business center of the region, is slowly turning its back on a fabulous post-war boom...the city looks to natural gas, cheap power and tourism....
> More than 700 lumbering operations, which blossomed because of lucrative American interest in the last 10 years, have sky-rocketed a snug 3,000 population into a housing short 12,000!...

In 1953 Western Plywood (now Weldwood) got a Tree Farm Licence that provided the company with an annual cut of 9.5 million board feet a year. Then a half dozen other applications for TFLs came in. Vanderhoof, Hazelton, and points in between were booming. Soon there were more mills between Prince Rupert and Endako than in the area farther east toward Prince George. This was the last era of genuine competition among companies rushing into unclaimed areas. It's an era myths have been built on.

One common myth, often heard at lumber manufacturers' conventions and in company submissions to government, is that competition meant speculation, and speculation drove out the poor but honest operator.

In his 1956 Royal Commission Report, Judge Sloan commented on this:

> *Competition*—A sale having been "put up" by an applicant and having been cruised, appraised, and advertised by the Service is then sold at public auction.... It is at this stage of the proceedings that competitive forces come into play. I have, in my opening observations...made a brief reference to the relatively minor degree the completely independent logger, now comparatively few in number, can influence the bidding. The real battles, when they do occur in the auction room, are joined between operators of bigger calibre, although perhaps represented in some instances by contract or tied loggers....

A "tied" logger is a contractor who is independent but who does all his business with one company and is tied to their fortunes, good and bad. Often the large firms help bankroll tied loggers, either by guaranteeing their loans at the bank, or through direct loans. Many allegedly independent contractors fall into this category today.

Back in the days of "free enterprise" competition for timber, a guy with a chain saw and six strong sons didn't have a chance to get good timber, as Judge Sloan pointed out. Even back then, people weren't just bidding on a stand of timber and the chance to cut it. Each sale implied the opportunity to get another stand later on—the "quota" system. One of the standard rules of the real world is that if competition is unlimited, the guy with the most money can drive out enough of his competition to make more money, then drive the rest out. Standard Oil is a classic case from the turn of the century.

Of course, some of the smaller firms were actually competing. In the Forest Service office where auctions were held, "there was fear, there were a lot of guys sweating. In the beginning there really was competition," according to the late Bill Grainger.

But to get away from this uncertainty, the operators in each Public Sustained Yield Unit banded together. In the Prince George Area the Naver Forest Association, the Crooked River Forest Association, the Westlake Forest Association, and the Willow River Forest Association were operating. Each of the Forest Associations, originally called Forest Protective Associations, had as members the operators holding rights in a particular Public Sustained Yield Unit. The Forest Associations had two purposes: to pay for surveys and for forestry work because the Forest Service didn't have the budget to do the work. And they helped keep out other operators.

Throughout the Interior, the Forest Associations have lasted. In 1978 the Willow-Abhou Forest Association included Rustad Brothers, Carrier Lumber, West Fraser, Netherlands Overseas Mills, and Weldwood as members, for example.

This group of operators logged in other areas as well, of course. Thus, the members of the Willow-Abhou knew what the members of the Takla Association were doing, since Netherlands was a member of both. The other members of Takla—Apollo Forest Products, Silvican Resources, Takla Forest Products, and North Central Plywood—also knew what was going on in the Nechako Forest Association, of which North Central Plywood and Takla Forest Products were members. The overlapping memberships in the associations end up, on an organization chart, like links in a chain. In addition to coping with problems in each area—such as which company would cut how much of the undesirable wood that had been hit by spruce bark beetles—the associations also made all the firms in the north interdependent. Pressure could be brought by association members against a mill that defied the wisdom of the industry as a whole. A firm could have its timber bid up at an auction, for example. Or it could be cut off from technological developments of forestry techniques developed by other members. As one millowner said: "If you develop something that gives you an edge, you've got to let everyone see it, like it or not. That's the rules of the game."

The associations divied up the allowable cut, deciding who could bid and who would bid on individual blocks of timber, in violation of an unenforced section of the Forest Act. But that isn't why the associations came into being.

Larry DeGrace, an ex-Forest Service member who started the Industrial Forestry Service, a consulting firm in Prince George, realized that government didn't have the staff to do a lot of things that had to be done. He reasoned that the only way to get them done was to have the operators who were in an area—a PSYU—work together in concert, paying for the necessary forest surveys and replanting. He was right. When the role of the Forest Associations was discussed with five different members of the Forest Service in Victoria, they all said the same thing:

> *We couldn't have survived without them. There were times when we called Larry DeGrace to find out about areas we hadn't been in. Our staff in 1953 was 101 for the whole district, then down to 95 the next year. And a lot of those were clerks and typists. Our people never had enough time to do the job they were supposed to. We've always found the Forest Associations very helpful.*

As recently as 1980, a Forest Association hired an expert on beetle kill to work with the Forest Service on a realistic strategy for coping with the problem of infested timber. In addition, the association worked out who would cut what. Since beetle-killed wood can be hard, dry, and difficult to saw, the "opportunity" to cut meant only that a firm wouldn't be able to cut elsewhere. The association handled the problem—and helped control the problem.

Chief Forester Bill Young spoke of the days when he was District Forester in Prince George:

> *I was always glad the associations were there. There are always conflicts between companies about the areas they operate in. Then we, in the Forest Service, are in the middle and we get lobbied. It's very difficult to make a factual decision. But with the associations a consensus is reached. Conflicts are behind doors and then we in the Forest Service only have to judge on the forestry questions involved.*

Another myth about the "era of competition" is that competition was so cutthroat that it almost became outright war. This argument has been used before by Royal Commissions, and by the industry, to explain why competition for timber rights is bad. But in 1952-56, at least eighty-nine percent of all auctions were held without any competition at all each and every year, and in 1952-58 the average number of competitive auctions was less than six percent.

Like the $50,000 chip plant announced in the Prince George *Citizen* at the end of September 1955 that was supposed to supply pulp mills, the practice of fixed bids never saw the light of day.

Meanwhile, lumber production kept going up, from under 500,000 board feet in the Interior—the entire area east of the Cascade mountains—in 1946 to 2.2 billion board feet in 1955. That year the lumber cut in the Prince George District rose forty percent to over 651 million board feet. The estimated financial return to the mills was $44,308,000, with almost three-fourths of it going to the mills on the East Line of the CNR between Prince George and the Alberta border.

The next year the annual mill list showed 801 mills, with 2,316 logging operations supporting them. Although the price for rough lumber dropped, the price paid for timber went up. This rise wasn't because of competition, but because the average number of acres per sale jumped from 172,188 to 208,970.

The annual report for the district noted:

> *It is interesting to note that although the cut increased to a record high, the number of operating mills dropped. The apparent discrepancy is explained by the increase in the number of stationary mills in the district. The new mills in*

effect replace several of the portable or semi-portable mills, in that their daily capacity is higher.

Then it went on to give a backhanded compliment to the Forest Associations:

> *Although the District has in the past had almost complete RCAF photo coverage available, we have not been able to keep photo coverage up-to-date....During the past year we received authority to purchase 26,000 photos from the new flying available....For the first time in years the department is almost as well equipped with air photos as are the industrial foresters.*

However, the big events of the year—like Cornell Sawmills at Dewey selling out to Penny Spruce Mills after twenty years in the business—were wiped out by a single headline. Robert Sommers, Minister of Forests, had been accused of impropriety in giving out Tree Farm Licences, and now he had been replaced.

In 1955 Gordon Gibson, the "logger-millionaire MLA," accused Sommers of having been influenced. Then Charles Eversfield, bookkeeper for H. Wilson Gray, a timber operator and head of Big Bend Lumber Company, went to a lawyer and made a complaint. The lawyer went to the Attorney General, Robert Bonner, and accused Sommers of accepting $7,000 in loans.

After months of political debate, Robert Bonner turned the matter over to the RCMP in February 1956. Throughout the spring and summer the government, and Sommers, avoided discussing whether or not he had actually received $7,000, some furniture and rugs, and a free trip to Detroit for his daughter's wedding, plus free lunches at the Terminal City Club.

The Premier called an election for September 19.

Bonner received the RCMP report in the summer, after the election had been called.

Sommers was renominated in Rossland-Trail, then the government was re-elected, including Sommers.

During the debate on the budget vote for the Minister of Forests, the questioning went on for days. At this point a Minister of the Crown can be asked anything about his portfolio that the Opposition wants to bring up. But Sommers didn't answer.

Here is Ray Williston's version of what happened:

> *It became obvious he wasn't going to answer. W.A.C. Bennett dissolved the house at 3:40. He grabbed Ken Kiernen, Bob and me. He demanded Sommers' resignation at 5:00. I was sworn in at 8:00 and that night I was doing his estimates in the house—until eight in the morning the next day.*

The Sommers case slowly proceeded through the courts to the Supreme Court of Canada. Finally Sommers pleaded illness and let matters lapse. On October 31, 1957, a commission of inquiry was appointed, headed by Chief Justice Sloan. It sat for a half-hour, looked over subpoenaed evidence, and concluded that a criminal prosecution was reasonable. On November 21, Robert Sommers was arrested. The case ran from May through mid-November 1958. Robert Sommers was convicted.

By the time he was convicted, his successor, Ray Williston, had caused a revolution in the forest industry in the north, creating the biggest boom of all.

Chapter 14
Farce: The Third Royal Commission

The Sommers case gave the public the impression that the government was full of crooks and that the forest was being mismanaged. To clear the air, and to give Williston the appearance of carefully taking counsel on how the forest should be managed, Judge Sloan, Chief Justice of British Columbia and author of the 1945 *Report of the Commissioner Relating to the Forest Resources of British Columbia,* was asked to look at everything from forestry education to the control of moisture run-off. The only reference to the Sommers affair was point eleven of the thirteen tortured sentences in his mandate: "Acquisition of rights to forest lands and timber...and the extent to which adequate and proper exercise of the rights thereunder is now made."

With this mandate, the man who had decided in a half-hour hearing that Sommers should stand trial went to work defending his 1945 report, attacking his critics, providing updated statistics on the industry, and doing everything *except* discuss the Sommers case or the Ministry of Forests.

Forest Management Licences—now called Tree Farm Licences—were discussed for over seventy pages. But only three paragraphs dealt with Sommers or his handling of the British Columbia Forest Products TFL application that was the backbone of the Sommers Case. The man who had suggested this type of licence in the first place wasn't going to say his recommended structure had failed to prevent abuse. His 1956 report didn't make a single suggestion that there might have been governmental abuses, or that safeguards should have been devised.

Instead, there was a lot of discussion about the need to cut down on waste, the need to expand operations in the Interior, the need for better utilization, and a lot of excerpts of dialogue between Sloan and Dr. Orchard, the Chief Forester. When not boring the reader with tables of figures, Sloan put in statements like this:

> *This design must be more than a mere forecast: it must fashion and mould the future. In that larger and creative sense, the paramount and guiding consideration must be the welfare of the people of this Province as a whole, an objective transcending in importance the future of individual industrial units, large or small. What we plan now is, in its larger and comprehensive aspects of principle, irrevocable to the extent that the plan is carried into operation. In this extension into time of present planning there comes a point from, and at which, there can be no return and no rational departure....*

Except by closing the book right there at page fifteen. It is not recommended that anyone read the 1956 Sloan report while sober. Or drunk. The next Royal Commissioner, Peter Pearse, tossed off the report with two mentions that it existed.

The report did, however, give the government a clean bill of health. It let Williston use a few of the hundreds of proposals scattered through the two volumes totalling 863 pages to do what he wanted to do. Sloan was elevated to the position of Forestry Advisor to the Minister, but in his new post he never made a single recommendation.

Chapter 15
Genesis

Between 1952 and 1965 new towns grew in the woods. A new class of millionaires was created by the provincial government. In the first decade the amount of sawlogs cut in the Prince George Forest District went from under 3 million cubic feet to over 151 million. Wood that had been ignored for decades was the object of fierce bidding at auctions. Trees "too small to bring a profit" became the major source of profit, and wood waste that had been burned for years changed into a cash commodity that kept mills alive during tough times. A pulp mill economy was superimposed on an improved sawmill base in a monumental economic development program that populated the north. It was the biggest change in British Columbia since the Canadian Pacific linked the province with the rest of Canada.

Although the story has been told many times, usually only a small part of the story is told. Here's the whole story, showing the myths for what they are.

Ray Williston, who was Minister of Lands and Forests from 1956 to 1972, was the man responsible for most of the changes. He gives the impression of being an Irish horse trader: white haired, genial, but canny. When Ray was interviewed in October 1978 and March 1981 about what happened and why, he told the story with relish and with a careful attention to detail.

> *When I took over, I sure wasn't happy with the standards of utilization, so early on—at the time of trying to get some sustained yield going—I got close utilization going. I had to get a pulp mill to use the chips sustained yield provided. Ian Mahood helped me on that.*

Though sustained yield had been official government policy since 1947, and a number of Public Sustained Yield Units had been created, few nurseries produced seedlings to go on logged-over land, and little funding had been set aside for any aspects of forestry. Even requests for money to build roads to allow logging, which would provide the government with revenue, were difficult to get approved.

Suddenly new information and a variety of pressures combined to let Ray Williston change the situation.

In a submission to the Royal Commission on the BCR in 1978, entitled *International Relationships with Their Responsive Attitudes and Actions*, Williston said, in part:

> *A new interest developed in the occupation of lands throughout the northern hemisphere. Most governments had a northern programme for settlement, to balance Russian activity. The Canadian Federal government, then under Conservative leadership, formulated a comprehensive plan for improving conditions in the north. Provincial governments responded to the interest by producing their own plans for the development of lands which had identifiable resources for exploitation....*

Apparently the Soviet Union began a massive northern settlement plan in the 1950s. The United States was concerned because, at the time, over eighty percent of Canada's population lived within fifty miles of the U.S. border. The Americans wanted Canada to develop a northern program so that the land mass would be occupied. Soon afterwards, John Diefenbaker was elected Prime Minister, partly because of his commitment to the "Northern Dream." Almost every provincial government came up with a plan to help populate the north. British Columbia, for instance, moved to develop the forest industry in the north, more than quadrupling the northern population in fifteen years.

A lot of people knew the story, but nobody talked about it. Even today American interest in Canada's northern development, and how the BCR's plans for an extension to Alaska affect U.S. defense, isn't discussed.

This does *not* mean that the forest industry in the north underwent a revolution simply because the United States was concerned about a Russian settlement program. But American interest, and the popularity of Diefenbaker's Northern Dream in Canada, certainly helped convince other politicians that Williston had a good idea.

Convincing businessmen was made at least a little easier by an increased demand, throughout the world, for paper—which of course meant a demand for pulp. Paperback books, computer punch cards, and increases in magazine and newspaper circulation are just a few examples. At the time it was already obvious that the advent of photocopying machines would kill more trees than forest fires would. With increased demand came higher prices. And if a company felt that these higher prices would continue, or improve, it changed the economics of the situation. High prices made longer chip hauls economical.

Then came more information. In the late 1950s, the Forest Service completed its first inventory north of the CNR line. Improved inventory data on other areas were compiled.

"Looking at it, some things were obvious," Ray Williston said. "If we set tighter standards, we'd get this much timber"—he drew a rectangle on a memo pad. "Now, we'd been taking out this much"—he colored in about a third of the space he'd marked off. "We told the guys that if they went to smaller logs, we'd give them this much extra"—he colored in another third. "And since that had previously been considered only good for pulp mills, we'd give it to them at a pulp price: 55 cents for twelve years—and what was left over was this empty part," he said, pointing at the white space remaining, "this third band. They got the right to that by performance. We used the carrot instead of the stick."

This explanation of "third band" timber, almost identical in 1978 to the one he gave Pat Carney when the idea was new, is very important—and can be very confusing.

People who had bid on timber, and cut it to the existing standards of the time, used the timber in the first band Ray drew on his memo pad.

He offered them the chance to get more timber without paying for it at auction. Any mill that cut 25,000 feet, board measure, per shift and put in a barker (since pulp mills don't want bark) and a chipper (to supply the waste wood to potential pulp mills in a form they could use), that agreed to cut and process wood down to a four-inch top in the Interior and down to a six-inch top on the Coast, that agreed to use all trees over seven inches wide at breast height or larger in the Interior and over nine inches on the Coast, and that cut the stump within a foot of the ground was given one-third more wood than had been originally bid on. These standards were a lot tougher than the earlier ones, which allowed stumps to be eighteen inches high, permitted tops left behind to be anything under eight inches, and let loggers ignore trees under eleven inches wide, chest high.

The "third band" wood—and today that's its official title—was any wood left over after these standards were met.

All this may sound as unimportant as the price of pears in Patagonia, but it was what got pulp mills going in the north. Because sawmills suddenly could get more wood without bidding at auctions, and could pay very low prices for it, they began to harvest the wood. Now there was no need for pulpwood harvesting rights to be given away, since the mills would do it for the pulp companies. And once the wood was harvested, there was more money to be made from two-by-fours than from chips—especially before the pulp mills actually arrived—so lumber companies started finding ways to use this "smallwood" economically, to produce high-value lumber products. With an assured supply of wood chips, Williston's job of convincing companies to build pulp mills was made easier.

According to the man responsible for this idea:

> *Our aim in all this was this—I believe the people in the industry you have to think about are the skilled workmen. They have the right to expect continuity of employment. The idea was to maintain stability in the area. I was opposed to competitive bidding, but wanted to move utilization standards so that you paid for utilization levels whether you reached them or not.*
>
> *It was Ketchum and Ernst who made some other guys go. When we opened up the small wood, in the pulp-harvesting reserves, I told the millowners they were the stupidest guys in the world if they let the pulp mills get into roundwood. But they did. The first guy to sign an agreement to deliver chips was the last to deliver them. A good guy, but boy he was slow....*

Williston points out that the idea of close utilization didn't suddenly appear in his head one morning while he was putting on his socks. Tom Wright, Dean of Forestry at the University of British Columbia, came to his office and said, "What volume of wood from a given acre in the forest do you put in the boxcars up there in Prince George?"

"Gosh, I have no idea."

Wright had just finished a study showing that only twenty-five percent of the wood on the given acre reached the boxcar. The rest was left in the forest, broken, ignored, or sent up in smoke in a burner. "That meeting made the most powerful impression on me of anything that happened while I was Minister," Williston said.

The Minister then went to Harold Moffat, who had hired Ray as Supervising Principal of Schools just after World War II and who had been his campaign manager. At the time Moffat was the Industrial Commissioner on the Prince George Board of Trade. He responded to Williston's request for a study to verify Wright's figures by raising $200 and going to Larry DeGrace of the Industrial Forestry Service in town. The study DeGrace produced, when tabled in the legislature in the mid-fifties, formed the backbone of the government's forest policy on close utilization.

Getting the forest industry to use wood it had ignored and laying the groundwork for the pulp mills was, according to one opposing politician, "an example of Williston's genius at improvisation," but at the time this was just one of many schemes in progress. Another grabbed far more headlines.

Voters were demanding more and more government services, as is normal, without the desire to pay for them, as politicians also consider normal. Among other things, health care, dam construction, the government-owned railway, and new roads all pressed for more money; yet no one wanted higher taxes.

At this auspicious moment Axel Wenner-Gren showed up. Wenner-Gren, a Swedish industrialist, had made fortunes as the head of Electrolux vacuums and Servel refrigerators. He owned canneries, mines, and Svenska Cellulosa AB, Sweden's largest pulp producer. He had the cash to fulfill any dream.

In November 1956 Bernard Gore and Birger Strid, two of Wenner-Gren's associates, met with Premier Bennett in Victoria. They proposed to build a freight-carrying monorail system along the Rocky Mountain Trench north of Prince George that would whiz goods at nearly 200 miles per hour. This was the keystone of a plan that included pulp mills, sawmills, towns, hospitals, schools, and a large dam to provide electricity to power the monorail.

The result of the meeting was a letter of intent signed by W.A.C. Bennett and Ray Williston. According to the government, all the letter did was let Wenner-Gren spend $5 million doing studies and committed the government to examine the possibilities. According to the newspapers of the day, Wenner-Gren had been given timber and mineral reserves covering 40,000 square miles—nearly ten percent of the province. The newspapers pointed out that Wenner-Gren had indulged in similar schemes in other countries, but they'd never become realities.

As the government became more involved with the needs and possibilities in Wenner-Gren's scheme, power generation became more important, leading to construction of the W.A.C. Bennett Dam. Wenner-Gren got as far as starting a new forest company and beginning the instant town of Mackenzie, north of Prince George. But these moves took time. The government wanted something that would both make it look good and take people's minds off the Sommers affair, which was still before the courts.

It was hoped that Wenner-Gren would help. But Williston didn't wait to see. He continued to change the forest industry, keeping as much out of legislation as possible. Then, if something didn't work, or if it took a while to get going, he could make changes. The process was totally under his control.

Chapter 16
Exodus: Competition and Small Firms Leave the Forest Industry

Every year from the late thirties until 1957 the number of firms in the north went up as more people entered the industry. Then the number began to drop until, in 1971, only 135 mills in the Prince George Forest District produced lumber, down from 704 in 1957. But during this period the size of areas lumber firms had the right to cut in grew until some were larger than Scotland.

Ray Williston eliminated competition for timber rights and aided takeovers as a deliberate policy. But, from his point of view, the change was inevitable.

"Sustained yield" was interpreted by the government as early as 1948 to mean an equal amount of timber every year, forever. To reach this goal, government officials gave themselves the power to decide who should cut what, where, and when. It's best to ignore complexities like the Hanzlick formula, which the Forest Service used in its calculations. (This formula assumed that trees stopped growing long before they started to decay and that neither species nor technology would ever change in popularity. Aside from that, the formula was logical.) The Forest Service interpretation worked like this: if a forest takes 100 years to reach maturity—defined as when tree growth slows and the tree is still healthy—then you could harvest one-hundredth of the trees each year. This way you'd never run out of timber, as long as new forests started right on schedule.

But this formula means that there's a limit to what can be cut. The Forest Service tried various methods of saying who could get the available timber and developed an elegantly simple system that was totally outside legislation. It wasn't against the law. It simply wasn't covered by any legislation.

This was the "quota" system. It gave priority to operators who had regularly bid on timber in the past and who had performed as promised. The Forest Service let them get roughly equal amounts of timber—plus "close utilization" increments—if they cut 25,000 board feet per shift and put in barkers and chippers. "If they could not reach this point because of wood or capital, there was no future," Ray Williston said. "An old, secondhand sawmill was worthless. That is why quota was allowed to be sold."

In overcommitted Public Sustained Yield Units, the companies were allowed a percentage of the available timber equal to the percentage that they'd cut in the past. Then the idea spread to mean a set amount of timber per year.

When new Public Sustained Yield Units were created, as fifteen were between 1956 and 1961, a successful bid at the first sale meant you could be assured of timber ever after. And, in the five older units in the area, buying out a competitor meant you'd get his wood supply.

One reason for allowing trading in quota was that it got rid of the small, inefficient operators who were big enough to get quota in the first place but not big enough to expand. Once quota rights were established, the small operators could sell out and get far more money than an old sawmill was worth. In effect, this was a reward for helping start the industry in the north. And if a millowner was ambitious, once quota was established, he could go to the bank, borrow money to buy cutting rights, then expand. Some men made millions selling out. Others expanded and made millions.

Quota was an invented asset, created by the government and controlled by the Minister of Forests. He made the rules, and he had to approve transfers of timber rights.

Of course some people didn't think this was a good idea. People called quota "an unearned profit" and said that the government was giving away its timber for eternity. But companies realized, even as they rushed to tie up quota, that they were bidding on something that could disappear. It was totally a matter of Ministerial discretion. And that meant it could be taken away. The Minister said it would be taken away unless they performed. "This placed special emphasis on plants," Ray Williston said. To keep quota, companies had to modernize mills, then modernize again and again, going for higher speed and greater efficiency.

If the government gave you something worth more money than you paid—as happened when quota was first started (although some operators didn't realize it and sold out cheaply at first)—or if you paid hard cash for something the government could wipe out without even holding a cabinet meeting, it would be logical for you to support the government. The forest industry was logical. Further, Ray Williston understood the industry in a way that no Minister before, or after, ever did. It's obvious from talking to him that he genuinely liked the people in the industry. So people in the forest companies felt comfortable with the government and supported it. Reportedly, hundreds of thousands of dollars a year went to the Socreds as a party. This wasn't a bribe; it was the result of the industry's feeling that Ray Williston understood and was trying to help. "I do not expect you to believe me," Williston commented on this subject, "but I never saw $1 change hands on a forest deal for political purposes." If one talks with Ray, and with the heads of most of the forest companies in British Columbia, it becomes obvious that contributions to his party were the result of winning the industry's support rather than fear of losing quota.

This was all good for the government of the day. But the question arose whether it was good for government as a rule. By allowing trading in quota, the government didn't make extra money; in fact, Williston's successor, Bob Williams, felt that the government wasn't getting the full value for its resources. How could the government get in on the action?

"It's easy to criticize a lot of things I did, today, with 20-20 hindsight," Williston said. "But when you try to figure out a solution to the problem, what would you have done? How would you have forced modernization and hurt as few people as possible?"

The most often criticized aspect of the quota system, as it evolved under Williston, is the way that it eliminated competition. However, Williston commented:

> *Once we started to regulate the cut, we were regulating the industry. It's like being a little bit pregnant. You had to regulate the whole thing.... As I said, the day you set sustained yield and working to a volume rather than an acre*

basis, you're a managed industry. They are told what they can cut and the day's going to come when they'll be told how to use it.... I thought economics would close off the smaller operators because they couldn't find wood at their utilization levels. We left the Special Sale Areas to accommodate these people. We figured they'd eventually run out of stands suitable for the level of utilization they could go. But the really small mills could find really small patches. They've kept going a lot longer than expected.

To a few people, this attitude was shocking. However, it merely reflected real-life economics. Back in 1933 Edwin H. Chamberlin of Harvard and Joan Robinson of Cambridge demonstrated that "perfect competition" where a lot of small firms compete is very rare. More usual is a small number of firms that consider the actions and reactions of the rest of the group whenever a decision is to be made. For decades the forest industry talked about "free enterprise" but never explained that their version didn't have competition except in product markets.

To avoid competition was only logical. If unrestricted competition had actually taken place, the price of timber could have been driven so high—as firms tried to squeeze each other out—that only the rich would have survived. The alternative was to avoid competition and survive.

Despite operators' claims that during the fifties, in the "era of competition," there had been a "war" and that competition almost destroyed the industry, it didn't happen that way. Here are two overlapping views of sales in the Prince George Forest District from 1952 through 1958.

FROM THE SLOAN REPORT (page 164)

Year	Total Sales	No Competition Number	Percent	Competition Number	Percent
1952	478	463	97	15	3
1953	497	467	94	30	6
1954	478	447	93	31	7
1955	662	610	92	52	8
1956	623	557	89	66	11

This report makes it look as though competition was slowly increasing every year. But the Prince George District annual management reports use slightly different data, presented a different way:

Year	Total Sales	Direct*	At Upset Price	Competition Number	Percent	No Bid**
1956	627	138	431	68	10.8	31
1957	468	105	340	23	4.9	52
1958	328***	80	229	19	5.7	34

* Direct: sold without possible competition.
** No bid: areas surveyed by companies and put up for auction by them, but without any bidders at the actual auction.
*** Includes nine direct sales not taken up, fifty-seven cash sales in addition for things like timber salvage and tiny acreages.

Justice Sloan concluded that eleven percent competition in 1956 was not a threat to companies in the industry or to their security of operation. But the percentage of sales with competition, looked at in isolation, is misleading. In 1955 the average sale was for 266.5 acres and brought the government $9,104.

In 1956 the area increased twenty-five percent to 333.28 acres, but the price went up over ninety-seven percent to $17,973. Again, these averages—by themselves—are misleading. There wasn't much competition, but when it occurred it was intense. In the sixty-eight sales with competition, prices went up as much as eleven hundred percent. Fir ranged from $1.80 to $20.00; spruce went from $2.30 to $12.40, depending on the sale; and lodgepole pine had a low of $1.40 and a high of $13.00.

Why the range? "I was going to bid on some timber," an operator said, "and outside the door was a guy who had a mill. 'For $5,000 I won't bid against you,' he told me. I didn't pay him. But a lot of people did." Other operators have verified the story.

A commonly heard story is about a firm that was about to sell out to a larger one. The big firm, under its licence agreements, was supposed to stay out of a certain area. But under those agreements, the rules didn't say anything against buying out a firm whose major operations were in their "area of interest." So the big firm paid the small one to bid on timber in a certain PSYU. The operator then sold out to the bigger company and made an extra $100,000.

In order to find out what went on, the last three Ministers of Forests were asked, individually, "Did you ever have a case of collusion or bid fixing brought to your attention? Did you ever prosecute?"

> Ray Williston: *No, I never had to get closer than a threat.*
> Bob Williams: *No, never. Not one single case was ever brought to my attention.*

When Tom Waterland was interviewed, Bob Wood, who headed the group that drafted the 1978 Forest Act, was also present since it was soon after the bill became law and it was expected that technical questions would be asked.

> Tom Waterland: *I'm not going to answer that question.*
> Bob Wood: *That should be answer enough. It tells you what you need to know.*

Since a lot can be done with tone of voice, and voice tone does not come across in print, Waterland and Wood were told: "I've heard, for example, that someone got $100,000 for bidding on timber and then turning it over to another company."

> Tom Waterland: *Which situation was that?*
>   *About two years ago. I really don't want to go into details....*

This reply implied that bringing up details would be, in effect, preferring criminal charges.

> Bob Wood: *I think I know the case you're talking about. That was really invited by the government. It was a knee-jerk situation.*
> Tom Waterland: *Oh yes, I think I do too. But in this case it was based on an alleged political reality that was not really a reality. Instead it was a political theory applied naively.*

Dozens of people in the industry have commented on gentlemen's agreements among bidders, although Forest Service personnel often profess ignorance of them. The law against collusion, the sections of the Forest Act against fixing of bids, has been repeatedly broken. The government, according to the office of the Attorney General, has never prosecuted a single case.

Breaking the law made a lot more sense to operators than following it.

The standard textbook on this situation is *Competition and Oligopsony in the Douglas Fir Lumber Industry* by Walter J. Mead. The key to Mead's ideas is in that tongue-twisting word "oligopsony," which means a situation where there's only a small group of buyers. According to Mead, the mills in the Pacific Northwestern states react the same as in an oligopoly—the control of a market by a small group of sellers, such as soap and oil companies. Mead found that in the Pacific Northwest, where the U.S. Forest Service has far tighter rules than exist in British Columbia, collusion was the norm. Even though some sales were set aside for small firms with fewer than 500 employees, about a third of these didn't have any competition in the auction room. And the smaller sales had far more lively bidding than the big ones. This meant that the small operators often paid about two-thirds more per tree than the major firms. Collusion was so common that one firm tried to sue another for breaking an illegal agreement. According to Mead, when quiet talks, implied understandings, or unwritten rules of the game fail to work—when several firms discover they're short of timber in an area, for example—only three possible scenarios can be followed:

1. Firms may feel that starting a bidding war could end up with being "rubbed out" by wealthier competition, so the result may be the same as if collusion existed. Why try to take timber away from a forest giant who can afford to bid every stand of timber you go after so high that you lose money even when you win?

2. If the firms are of equal size, or if the competitors can't hit back for other reasons, there may be highly competitive bidding until gentlemen's agreements are re-established.

3. Since high bidding, uneconomic as it may be in the short term, might force another firm out of business, or a least out of an area, companies may try to destroy their competition if they feel safe from retaliation.

To take this from the theoretical level, Six Strong Sons Logging would be foolish to start a bidding war against Multinational Mills Ltd., because Multinational Mills could afford whatever it took to either drive Six Strong Sons out of business or make the company realize the error of its ways. And, of course, Multinational Mills would be able to make sure Six Strong Sons Logging stayed the size Multinational wanted—usually just big enough to stay in business so that Multinational could talk about "diversity of size" and "free enterprise" and keep the government from stepping in. In contrast, if Six Strong Sons' only competitors were Gyppo Logging Ltd. and Fred's Timber and Truckfarm—both about the same size with the same ability (or inability) to get money from the bank—competition might make sense.

All of these possibilities existed at various times in the north. Some firms were driven out of business. Often midnight meetings weren't necessary since companies in balance with each other didn't want to start trouble. And at times there were genuine—if brief—periods of competition.

However, these periods of competition taught the operators that competition cost them money. Gentlemen's agreements became common. And some operators screamed to the government. When the screaming became loud enough, Williston acted.

In 1960 a new rule let operators in overcommitted areas demand sealed bids when they applied for Timber Licences, instead of forcing open auctions.

And the new rule let the firm that cruised an area and requested an auction of the rights to its timber match the highest competing bid, to make sure the firm that had paid for the survey won the sale. In 1961 this was extended to fully committed areas. In 1964 the government added a requirement for a deposit that went to the government whether a firm won or not. In 1976 Peter Pearse summed up the effect in his Royal Commission Report:

> *The combination of the sealed bid tender option, the matching bid privilege and the forfeitable bidding fee soon discouraged competition to the point where it now rarely, if ever occurs.... The competitive timber sale policy that had been followed for 50 years since the **Fulton Report** in 1910 had been all but totally undermined by 1964....*

Quota plus lack of competition equalled takeovers. As one veteran mill executive said, "There were no guarantees, since quota depended on the whim of the Minister...but there were lots of brown noses." In 1958 the number of mills in operation began declining.

MILL FIGURES FROM P.G. FOREST DISTRICT
ANNUAL MANAGEMENT REPORTS

	1955	1956	1957	1958	1959	1960	1961	1962	1971
Mills in operation	730	687	704	669	648	621	566	545	135
Nonproducing mills	92	114	167	167	197	160	199	200	221

Nonproducing mills often remained listed while companies left them to rust, where they sat, rather than selling them for scrap or paying to have them torn down. And while the number of mills went down, production increased, from 674,094,000 board feet in 1956 to over 470 billion in the early seventies.

Chapter 17
The Era of Consolidation

If a small firm bought another mill, and its quota, it could build a bigger mill that could handle the wood from both firms while employing fewer people than the two original mills. That meant more money. If a middle-sized firm bought one of equal size, it could build two mills, each designed to handle a slightly different size of log, instead of having just one mill trying to handle logs from twelve to forty-five inches wide. The mills would need less adjustment and break down less often than if they'd tried to handle everything. That meant more money. If a firm bought quota, its future was less risky since the company could count on getting an increased amount of wood forever.

So each millowner with some money tried to buy out every firm he could afford. In 1957 the number of mills producing lumber dropped from 801 to 762. The practice of clear-cutting large acreages became more popular, although only 5,085 acres were clear-cut while 45,249 acres were selectively logged. In many cases "selectively logged" was a euphemism. It often meant selecting the best trees and leaving the rest to supply seeds to fill in the empty spaces.

There were 7,770 people employed in the forest industry in the north, but only 2,463 worked nine months or longer. The next year, 1958, was a better year for employment; 8,246 people were working in the industry, and 3,274 worked at least nine months. The number of mills at work dropped again, but the amount of wood cut went up as new, larger mills were built. The annual management report commented about cutting areas, in part:

> *CARP SYU (Sustained Yield Unit): Mills presently established in Block 1 are capable of drastically overcutting the unit. No "operators' association" exists but to date bidding against each other at auction has not occurred. Mills are in short timber supply and annual requests always grossly exceed the available volume....*
> *NAVER PWC (Public Working Circle): Existing mills are capable of cutting well in excess of the allowable cut and the overcut in Unit 1 illustrates the trend. Unit 1 also exemplifies the trouble caused by a reduction in allowable cut (from 1,455 to 1,000).* *

In 1959 prices jumped to $66 per thousand feet for spruce and $72 for fir. The industry reacted by putting 5,272 men to work in the woods and another

---

* This parenthetical remark is part of original text.

3,500 sawing lumber. Over 81,000 acres were logged by 3,111 logging operations. If that sounds as though each logging "show" employed fewer than two men, remember that after an area was cut the crew would move on. The size of tractors and skidders was going up. According to the management report:

> *It is fortunate that the accent is now on various forms of clear-cutting, so that damage to residual stands has not increased proportionally.... Only a minor number of operations are now using a truck log haul, but this will be more prevalent in the next few years.*

At times the trucks were hauling up to five miles. Today, seventy-five-mile hauls are common.

Five new Tree Farm Licences were established, the first seedlings from nurseries were planted in the north, and the number of mills dropped again, this time to 731. But overcutting was becoming a very serious problem as the mills responded to the highest prices ever. The Stuart Lake PWC was overcut by twenty percent. In Block 1 of the Carp, the overcut was twenty percent again, so every operator in the area had to be cut back nine percent. The Naver, Block 2, "was grossly overcut," the management report commented. One and a half million cubic feet "of the total cut was from the Quesnel operation on the South boundary. This operator has completed his Naver operation and is moving to his Cottonwood PWC area...."

To keep up with the action, the Forest Service staff expanded from 95 in 1954 to 148 in 1955 and 225 in 1962. But with the increase in cutting, the Service had a difficult time keeping up.

With this rush to cut and with the consolidation in the industry, it would have been logical for some firms to try to bribe Forest Service officials. A bribed Forest Service staffer in the right position could influence what a firm paid for stumpage and for overcutting and perhaps influence cutting rights, if only by helping the mill operator discover the best areas to bid on.

But no member was bribed.

Bill Young, who was District Forester in Prince George during much of the boom, said when asked about this:

> *It may appear to the public that, "the Forest Service takes care of its own," but a number of people have been released from the Service. There were a variety of reasons, that I really don't want to go into—having booze in a vehicle was one, for an example—but to my knowledge no one has ever been associated with bribes. If you look at the forest services around the world, that's remarkable. When there is an allegation of wrongdoing we bring in the RCMP right away. And we've done that often. I personally have been involved in firing a dozen guys—never on bribery. We have a reputation as a straight organization, perhaps a military type of organization as you say. We therefore have policy carried out top to bottom. And I've never found any record, any trace, or even heard of anyone being bribed.*

If anyone had been bribed, everyone would know about it just because of the way the forest industry works in the north. But in literally hundreds of interviews, it's never been mentioned. There have allegedly been a few cases of attempted bribery. But when they were mentioned it was carefully pointed out that none were successful.

The Forest Service had its own problems. In 1957 the north had one of the wettest summers on record, and lumber prices were down. It was followed by the worst year of fires since the Forest Service was organized. In that year, 1958, mills were destroyed and towns were threatened throughout the north. The following year the growth in the industry was almost as fast as the proliferation of fires had been the year before.

But, from the government's point of view, operators were still wasting wood, even though a new industry—making utility poles from lodgepole pine—had started. In 1957 Northern Forest Products opened a new mill to treat poles for "weeping," or running sap. The total output for the district that year was 454 *miles* of the new type of poles.

It still wasn't good enough. Ray Williston began giving public talks about the need to shrink the minimum size of trees the operator had to cut. In response the Northern Interior Lumbermen's Association commissioned a forestry professor at the University of British Columbia to investigate the situation. In his report he suggested that the minimum size of wood to be cut should be increased. Along came Williston with his offer of more wood if the operators were willing to go to smaller minimum sizes. The report was never mentioned again.

This was only a sideshow. The Federal Housing Administration in the United States, the government agency that stood behind most mortgages given by builders, said that effective April 1, 1960, all framing lumber had to be graded and marked according to "recognized standards." And the standards used by the Northern Interior Lumbermen's Association didn't conform—or, to be exact, they were too loose and they were unrecognized. That could have been the end of the biggest market for northern lumber, since there were few trained graders in the north to do the job, even if the grading system had met U.S. standards. Luckily, help was on the way—from the IWA.

After the 1953 strike, the union had maintained a low profile and had done its best to be a responsible citizen. In the mills union stewards made sure the contracts were more than just wage agreements but were enforced every day. Management came to realize it had to live with the union, instead of just fighting it during negotiations. And both sides came to understand they had common interests. Bob Gallagher, Secretary Manager of the Northern Interior Lumbermen's Association, said of the new union president, Jake Holst, "Jake was the first person the union had who negotiated in private rather than through the newspapers."

When word came in 1958 that the grading standards could, in effect, be used to embargo B.C. lumber, both management and the union went to the States to lobby. The IWA supported management's grading program. When contract negotiations came up in 1958—while the grading system was still young enough to die unless nourished—attitudes were far different from attitudes during the strike. Bob Gallagher said:

> *Unions were becoming more respectable in the public's view, and more acceptable to management and more responsible. Both sides had realized they had to keep the industry going. And you have to remember the public wouldn't have tolerated a stubborn stand on our part if they thought it was stupid.*

The 1958 contract called for a union shop for all employees hired after September 1, 1956, raised the base pay rate to $1.69 an hour, and allowed rest periods for the first time.

In the fall of 1959, thanks to modifications in the grading standards used in the north, lobbying by the companies, lobbying by the B.C. union and by its U.S. counterpart, plus factors including an upturn in the market that removed some American firms' fears of Canadian competition and perhaps Williston's northern development plan, the grading standards used in northern British Columbia were accepted, with only minor modifications.

It was the last time exports to the United States were seriously threatened.

Chapter 18
The Smell of Money

Planer mills. Photo courtesy of B.C. Ministry of Forests.

When Ray Williston tells of getting the pulp mills established in the north, he makes it sound easy. It wasn't. He tried a variety of approaches that met with blank stares until he developed the Pulpwood Harvesting Agreement (PHA) in 1961. If a pulp mill was established in the area where the government wanted it, the Agreement guaranteed the right to use wood not normally used by sawmills: anything eight inches or less in diameter a foot from the ground. The rate was fixed at 55 cents per hundred cubic feet. On larger wood, normal stumpage rates had to be paid. This agreement was needed because although sawmills were wasting enough wood to supply several pulp operations, they weren't turning that waste into anything pulp mills could use. Anyway, pulp companies didn't trust the government figures enough to rely on mill waste when they were investing $60 million to $80 million.

Sawmill owners were not, of course, enthusiastic about the idea of letting other firms into their areas. To help avoid conflict, a rule was established prohibiting pulp companies with Pulpwood Harvesting Agreements from competing with sawmills at timber sales in PSYUs and forcing the pulp firms to purchase chips wherever "economically feasible."

The set rate for pulpwood helped encourage potential pulp investors since the rate was fixed and low. Sawmills responded by looking at how to use smallwood—and keep the pulp mills out of their areas.

Canadian Forest Products—by this time the employer of Tom Wright, who had first brought the waste problem to Williston's attention—was interested, as a result of lobbying efforts by Wright and John Liersch, then a vice-president of the firm. Poldi Bentley, cofounder of the firm with John Prentice, later recalled:

> *Up until that (PHA) legislation, there wasn't a single pulp mill in the Interior. Under the guidance of Mr. Liersch we could see a tremendous chance to build a pulp mill in the Interior and utilize the waste of the sawmill industry....*

It was quite a gamble. One firm, MacMillan Bloedel, had conducted a study concluding that a pulp mill would never be economically feasible in the north. Wright, Liersch, John Stokes, Larry DeGrace, Williston, and Poldi Bentley "himself, pipe and all, (were) all out there in the bush measuring the gol-darned trees to see whether it made any sense," Ray Williston said. It did. In addition to the low-cost wood in the forest, there was a surplus of wood residue from the mills, plus an increasing world demand for Scandinavian-style pulp. And because of the government's offer of a Pulpwood Harvesting Agreement, investment capital was available. The Prince George area met other criteria too. It had plenty of water, for instance. Abundance of water was essential since a ton of kraft pulp requires 35,000 to 55,000 gallons, mostly to wash the pulp during processing.

On May 2, 1962, the Prince George *Citizen* reported:

> Gov't Approval to be Sought
>
> ### $50 MILLION PULP MILL FOR CITY PLANNED BY CANFOR
>
> ...*Canadian Forest Products plans call for construction of a $50 million mill on the Fraser Flats just east of the city designed to produce 500 tons of fully-bleached kraft pulp sulphate per day....*
>
> *(John Liersch said,) "We are now in a position to present a formal proposal to the Minister of Lands and Forests for his consideration.*
>
> *"In compliance with provisions of the Forest Act, our proposal will provide for maximum economic utilization of both sawmill and logging waste in the area."*
>
> *Lands and Forests Minister Ray Williston told the Citizen during a recent visit here there would be no delay in arranging a public hearing if such an application were made by a "responsible firm."...*
>
> *Asked if the sulphur smell from the mill is likely to affect the city, Mr. Bentley said this would be minimal.*

A pulp mill close up smells like cabbage cooking in an outhouse. But whenever anyone complained to Premier W.A.C. Bennett, he'd smile and say, "That's the smell of money, my friend."

When the pulp mills started up, they stank. Today, after many millions have been spent on environmental control, they only stink sometimes, and then far less than they used to. But residents still keep their windows closed on foggy days.

On June 6 the hearings on the pulp proposal started. Canadian Forest Products wanted pulpwood rights in the Parsnip, Crooked River, Carp, Stuart Lake, Nechako, Westlake, Naver, Big Valley, and Willow River Sustained Yield Units. That's about eight million acres. And although Williston interpreted—and caused the companies to interpret—pulpwood as eight inches in diameter or less, as Peter Pearse pointed out in his Royal Commission Report in 1976:

> *The three earliest contracts convey an option over "pulpwood" defined as residual material "below sawmilling standards." When they were signed, the "intermediate utilization" standard under which sawmills were recovering timber meant that much of the material they did recover as sawtimber and many entire stands clearly fell "below sawmilling standards." But the shift to the "close utilization" standard...has largely absorbed that wood.*

Perhaps Canadian Forest Products could foretell the future in their request for so large an area.

They got a lot of what they'd asked for.

After the hearings, the Pulpwood Harvesting Agreement went through seventeen drafts before Williston and Canadian Forest Products hammered out a deal both sides could accept. The company wanted everything it could get since it was taking a big risk. Williston knew the Agreement would set a precedent.

Although each Pulpwood Harvesting Agreement differs from the others, the outlines remain the same. The company has to build a pulp mill of a specified size. In return, the government provides an option to purchase pulpwood, which is defined in varying ways, depending on when the agreement was written. The deal is for twenty-one years. All wood harvested either has to be used for pulp or traded to a sawmilling firm for pulp logs. This clause gave every Interior pulp company the idea of setting up another firm in the sawmill business. In the case of Canadian Forest Products, the mill that was built, called Prince George Pulp, is not part of Canadian Forest Products but is owned by Canfor Investments, which has the same ownership as Canadian Forest Products. The sawmill side is another Canfor Investments firm too, Takla Logging Ltd. To avoid confusion, the company negotiating will be referred to as Canadian Forest Products and the mill as Prince George Pulp.

Canadian Forest Products agreed to use as much sawmill waste as was economically possible, and, like most other Interior pulp mills, it received an understanding from the government that the firm would have the exclusive right to purchase chips from other firms in the area covered by the Pulpwood Harvesting Agreement. This right kept chip prices artificially low for almost twenty years. The next several pages of the agreement said that the firm had to put up a performance bond and that the government could make changes.

The only detail not spelled out in this Agreement, or in any of the others the government signed, was the understanding that each Agreement defined an exclusive area where no one else could buy chips.

Instead this understanding sometimes showed up in statements on "chip direction policy" and in contracts with sawmills that put in chipping equipment to qualify for third band timber. In the agreement with the lumber producer there was often a clause saying who would get the chips produced.

When Canadian Forest Products got the deal with the government, it had a partner in Prince George Pulp, the Reed group from England, which Canfor Investments later bought out. Poldi Bentley commented:

> *The reason, in part, why the venture was joint was due to the ground rules laid down by the government. They said you have to be able to arrange financing and have a market, and then you can get the timber allocation. At the time fulfilling these conditions required one to have a connection with one or two large buyers.*

Reed was a very quiet, but very active, partner. The industry was convinced that Canadian Forest Products was going to welsh on its agreement. Ray Williston tells the story like this:

> *You have to remember MacMillan Bloedel had said Prince George had no potential for pulp at all. None of the major firms believed it would happen. I kept talking about Canfor proceeding by such and such a date. A representative from one of the major firms came over to my office and told me I was about to be embarrassed, that nothing was going to happen. "We have contacts with every office that deals in pulp equipment," he said. "There's not one line on paper, not one contract let. Every time John Liersch"*—who was in

> *charge of the project for Canfor—"goes anywhere, we know who he's talked to, where he's been."*
>
> *What they didn't realize was that the Reed Corporation in England, which had been buying pulp from MacMillan Bloedel, said they wanted a more Scandinavian-type pulp. MacMillan Bloedel told them to buzz off. So Reed and Canfor went together to build. The majors hadn't bothered to check with Reed, and they'd had Prince George Pulp designed in Dryden, Ontario, and in their London offices. In fact, they'd secretly had an entire shipment of logs sent from Prince George to Dryden so they could compare the pulp from our trees with that from Scandinavia. They found them alike.*

Construction of the mill started in 1964 and was completed by April 1966, after an expenditure of $84 million—168 percent of the original budget.

Williston had thought that after the first agreement had been signed it would take up to five years before another one would be negotiated. This interval would provide time for administrative problems to be worked out. "However," he said, "the concept was accepted so rapidly that within the five-year period at least six projects were either built or had reached an advanced stage of negotiation...."

After Canadian Forest Products got its Pulpwood Harvesting Agreement in 1963, a Kamloops operation was approved. Northwood Mills was awarded a PHA in 1964. In 1965 Intercontinental Pulp—backed by Prince George Pulp and Paper, and Feldemuhle of Germany—got one. The same year Price, from eastern Canada, and Weldwood formed a joint venture called Cariboo Pulp and Paper and received a PHA in Quesnel. In 1965 still another PHA went to the Bowaters/Bathurst pulp companies, which were granted an Agreement whose area centered on Houston. The joint venture only got as far as a sawmill, then Northwood took over when the firm got into economic difficulties.

One of the most interesting deals of the era was a Pulp Harvesting Area that was designated for an agreement but was never awarded. Vic Brown and Al Farstad from the east Kootenays contacted Mitsubishi in Japan about the possibility of a pulp mill. They were put in touch with Honshu Pulp and Paper, then together formed a firm called Crestbrook Pulp and Paper, which applied for a pulp licence. Another firm in the area, Crows Nest Industries, didn't like the idea. According to Ray Williston:

> *The latter firm really admitted that it was not ready to start a pulp mill, but the principals were going to do all they could to prevent their competitors from getting started. I was left with no alternative but to hold a bonus bid auction to determine who would be awarded the licence. This was held in Cranbrook. When the bidding reached $1.00 per unit I called a halt because no one had yet worked out the economic return from such a venture in this area already known to be operationally more difficult than any of the others previously awarded. I sent back all of the bid deposits and told the parties to cool off and consider their positions. I do not have to tell you the fun that political opponents and political science professors had with that decision....*

For years the Crestbrook auction was used as an example of political interference in the "free market process." It was. It was also an excellent example of "grudge bidding"—one firm was out to hurt another rather than to get wood it needed.

In Mackenzie, another competitive auction took place, with different results. Alexandra Forest Products, started by Wenner-Gren, had been taken over by B.C. Forest Products. Robert Cattermole, an operator who had started on the Coast and gradually sold and bought until he was operating in the north, wanted to get into the area too. Another contender was Ben Ginter, who had once run a logging camp and small mill in the Prince George area before moving into road construction, brewing, and other ventures. B.C. Forest Products, Cattermole—in partnership with a Japanese firm in a joint venture called Finlay Forest Industries—and Ginter all showed up at the public hearing to decide who could have rights in the area.

Before any firm had approval, Cattermole had bulldozers at work clearing a mill site. Williston ended up approving both the B.C. Forest Products application and another application for the same area. Ginter and Cattermole bid against each other for this second Pulpwood Harvesting Agreement. Cattermole won, at a cost almost as high as the one that caused Williston to withdraw the area Crestbrook and Crows Nest had bid on. Finlay built a sawmill and a market refiner groundwood pulp mill.

This was the first groundwood mill in the province. The difference between groundwood mills and the other mills is simple: take a piece of wood and rub it against a rough stone. The result is wood that's ground into fibers—groundwood—and can be used for making paper.

Other pulp processes include using old rags, recycled newspaper, some grasses, and chemicals to reduce these raw materials to a mass of cellulose fibers. These wet fibers are then put on a screen, pressed, and dried. Kraft pulp, the most popular type in the north, is made by cooking wood chips in sodium hydroxide (caustic soda).

The whiteness and the strength of the paper depends on the length of the fibers, the species of wood, and the method used to make the wood chips become pulp. Kraft pulp is very white and is strong enough to be used to make the paper bags for holding cement. It also strengthens grocery bags and toilet paper. Although groundwood pulp starts with the same type of chips, it isn't as strong. However, it is a lot cheaper to make, so it's often used to add bulk, mixed in with other types of fibers.

Alexandra Forest Industries, Mackenzie, October 1968. Photo courtesy of B.C. Ministry of Forests.

Paper rolls at Prince George Pulp. Photo by Brock Gable, courtesy of the Prince George *Citizen*.

Cattermole sawmill, Mackenzie, Oct. 1968, with proposed mill site in foreground. Photo courtesy B.C. Forest Service.

This description makes it sound as though pulp mills are giant chemical labs controlled from rooms bursting with dials and buttons. To some extent they are. People are needed too, however. People operate the machinery in the plant and keep it running, people move the chips to the plant, people do quality control work—and when Mackenzie was started, there was no room for people. Mackenzie was just a clearing in the woods. In the beginning construction workers lived in tents at the end of a dirt track until quarters were built. When the first sawmill started operating, the men moved into the just-vacated construction quarters, but there wasn't enough room for everyone. A hotel was built and the men moved in, four to a room. They hated it. When houses first became available, there weren't enough to go around, and the men who didn't get them became more annoyed.

Mackenzie was an "instant town," and no instant town had ever, or has ever, worked in British Columbia.

In the 1950s the Forest Service had tried a few "loggers' subdivisions" in the bush so that the men could live near the mills. None worked out. Today most are either slums or weekend retreats, though a few of the houses have been turned into nice homes by people who don't have to go to work elsewhere.

You can't build a town overnight. Kitimat remains proof of what happens when you try—a typical topic of conversation is how many days are left on one's work contract before one can get out.

The planning of Mackenzie didn't allow the time needed to create a town. In the first years there was one store of each type, and the trip to Prince George took about three hours in summer and much longer in winter. Wives who moved to the town usually left—with or without their husbands. Women who had worked couldn't find jobs. There were no places for women to spend time outside their homes. The amenities city people and townspeople take for granted hadn't had time to be built. It wasn't a matter of bad planning; it was a matter of the priorities of the companies involved. The pay was excellent, but living conditions were bad enough to keep tempers short throughout the sixties and into the seventies. As one early resident said, "How they treated where we lived, and how they decided who got the houses, made it obvious they didn't give a damn about us." He left as soon as possible. So did a lot of others.

Other towns were having problems too. Houston faced a similar housing crunch, although it at least had a town center that helped provide a settled feeling. Prince George doubled in population again.

It may seem obvious, in retrospect, that there wouldn't be enough skilled tradesmen to build all the pulp mills and sawmills demanded by the boom. But at the time the only need seemed to be for more money to hire men away from other projects or to entice them to the west. The result was a floating community, composed largely of young men with money burning holes in their pockets. And where there's money, there will be land speculators, prostitutes, merchants who charge outrageous prices, bootleggers, and drug dealers. The undesirable elements rushed north.

The new pulp mills affected the lumber industry, of course. A firm could put in a barker and a chipper and get a higher quota and a new product from mill waste that had a guaranteed market, since a single pulp mill used at least fifty railcars of chips per day. In addition, the pulp mills—with the exception of Prince George Pulp—were usually connected with existing lumber companies, and it was obvious that by increasing the size of the lumber side of the business, payment for chips would simply be taking money out of one pocket and putting it into another. So the firms building pulp mills tried to buy out other lumber firms and applied for any timber that hadn't been committed to a company.

If a small firm wanted to get in on the action, it had to at least buy a barker and chipper so that the size of its quota would be increased. That meant an outlay of about $215,000. Because really small firms couldn't afford this expenditure, there was more consolidation of firms. The minimum size for economic efficiency kept going up, creating millionaires as larger and larger firms were bought out.

The arrival of the pulp mills changed the size, shape, and style of the forest industry in the north. Suddenly chip prices helped keep mills in business when lumber prices dropped. Since close utilization was needed to get the extra quota the government promised, and chips were purely a by-product at most mills, companies rebuilt their mills to handle the larger volumes of wood reaching their log yards. Today pulp dominates the forest industry in the north.

Here's why, according to Ray Williston:

> *My theory was that demand could only come after facilities had been built. No pulp mills—no use for chips nor any enforceable demand to make chips. My goal was the approval of enough pulp capacity to require the very poor sawlogs to be chipped.*

Ray Williston reached his goal, getting more use out of the wood on each acre of forest.

The process he went through to do this has been attacked, defended, and commented on more than anything else in the forest industry except perhaps sustained yield. There's good reason for this. Williston is the most important individual in the history of the forest industry in northern British Columbia. He made the industry what it is today—good, bad, and in between.

Chapter 19
Big Money

The boom in the north caused by close utilization and the coming of the pulp mills didn't unfold simply and logically. To the mill operators at the time, the future looked confusing, with various indicators pointing in different directions.

In 1960, for example, prices for lumber started strong and then slumped during spring break-up. They stayed low for the rest of the year. A lot of mills shut down. Some operated at a loss. At year end about 127 million feet of lumber was sitting in mill yards in the north, double the amount that had been there at the end of 1959. The number of mills listed dropped from 731 to 704, but only 621 operated during the year.

In 1959 Midway Terminals had bought Sinclair Spruce and Upper Fraser Mills, and now it was buying other small firms. Later the company changed its name to National Forest Products—just before it disappeared.

The year was a good one for grumbling. And it wasn't limited to the millowners. The first section of the Forest Service annual management report is headed, "Highlights of Management Activity." "I doubt if such actually existed in the past year," the report began. The section ended:

> And while this report is being compiled, the thought arises as to whether the written section of the report is necessary. The argument is advanced that these reports are a valuable reference and are used extensively in Victoria by other government departments. This I doubt very much....

The same cheerful tone ran through the report:

> Clerical: No need to comment on this, as the turnover and lack of replacement continues to dog the efficiency of the management section.
>
> Ranger staff: The question still arises as to why the necessity for all the paperwork now submitted and this office is inclined to agree....

Ralph Robbins's discussion of working plans clearly showed why the Forest Service staff felt they were being picked on:

> The biggest uproar of the year was the Department's decision to refuse to accept any applications in the Monkman S.Y.U. Block 1 of the Bowren and Block 1 of the Big Valley and instead fully commit by means of Forest

> *Service sales. Our working circle staff laid out 17 timber sales and to date only the seven in the Big Valley S.Y.U. have come to auction. Bitter competition ensued for these sales, the highest of which was bid from $2 to $9.00 per hundred cubic feet. We were advised the bidding was for future position and "quota" in the unit....*

The companies competing often talked about "trees for tomorrow," but their actions showed a different attitude. One section of the Carp was overcut by forty percent. All three administrative units in the Naver near Hixon were overcut, averaging twenty-seven percent in an area that had been overcut the previous year. And this was in a bad year for lumber.

A new Public Sustained Yield Unit was formed, south of Tete Jaune Cache, called the Canoe PSYU. It was established to help get rid of timber in the area that would be flooded when the Mica Creek dam was built. There were still plenty of logs left when the dam was built, and a lot of trees that were still standing drowned. Since this was a Public Sustained Yield Unit it had to include more than just the area to be flooded, and each year figures were published showing an allowable annual cut. But the area that wasn't flooded wasn't surveyed until 1974. Its almost 2,000 square miles of forest were just too low a priority to bother with on the Forest Service's limited budget.

The Canoe PSYU is an extreme example, but the limited budget caused problems elsewhere as well. The inventory staff didn't have enough manpower or money to keep up with the expansion of the forest industry, so growth factors were added to whatever had been found in the last survey. Because the possibility of pest infestations and the effects of warm winters or long ones were ignored, the data for much of the north were soon proved wrong. Near Longworth, for example, a decade passed between inspections. A spruce bark beetle epidemic spread for over five years before it was discovered. Considering the budget, this laxity wasn't surprising. What *was* surprising was how much was known about the forests.

The Forest Service was facing other problems too. In 1961 the turnover of the staff that cruised the timber was eighty-eight percent. There was so much paperwork that foresters had little time to see the forest. Rangers were working with new rules that caused problems—like the one that let the firm that was "supposed" to get a stand of timber match the bid of anyone who beat it at the auction. Competition was on its way out. The annual management report said:

> *With the...publication of the quota lists for all units, the Department, inadvertently perhaps, has now guaranteed each licensee a given volume and the sooner such fact is recognized by all concerned, the sooner will progress be made in our management program. The cut-cruise discrepancy has become the problem and it is respectfully suggested that the time has come to discard our inadequate cruising system and antiquated scaling farce. It is ridiculous to realize our policy rewards an operator with more timber because of his poor utilization and deliberate under-scaling.*

The farce was this: a company would ask for an auction of a stand of timber and an inventory. This "cruise" was done by, or under the control of, the Forest Service. It almost always underestimated the amount of wood in an area. "A guy would figure out the wood on one acre, multiply by the number of acres, then divide by two to allow for poor wood," a forester said, exaggerating a little, but only a little. "Anyone who couldn't get out forty percent more than was allegedly in the sale wasn't trying." This meant that an inefficient operator who took only the wood that exactly fitted his mill equipment would get just what the cruise said he would and wouldn't have to pay any more money. Then, of course, he could ask for more wood. The careful operator was penalized by the system.

When lumber prices slumped, as they did in 1961, the temptation to do this was great.

The year was so bad that only 566 mills in the district operated. The IWA signed up fourteen new mills, but membership dropped by twenty percent. One reason for declining membership was automation. Companies looked at equipment and discovered that hardware couldn't go on strike or ask for higher wages. So D8 Cats went in, hauling up to twenty trees at once, and that meant fewer men at work. New five-ton Fords and war surplus six-by-sixes were used to skid logs to bush mills. One operator had a tank retriever dragging in 1,500 cubic feet per load. The use of trucks to haul logs to mills centrally located in the larger towns, instead of moving the mills to the logs, also became popular. Truck sizes doubled and tripled into $60,000 investments, hauling logs thirty to seventy miles. Stationary bush mills put in edge trimmers to reduce the weight of the green lumber hauled to planers, and that meant more boards per load and fewer men again. Green chain sorting—where men stood in line and waited for the lumber to pass by, each stacking one or two types—began to be replaced with automatic sorters. That eliminated a lot of jobs.

A new plywood plant opened at Valemount and began testing cottonwood. National Forest Products got into financial difficulties. When it went into receivership, Noranda Mines Limited had just liquidated the Waite Amulet copper-gold mine in Quebec, and the company put some of the money into buying National.

"It seemed like a relatively small investment for us whereby we could get into another major resource sector in Canada, and so to speak dip our toes in the water," Adam Zimmerman of Noranda recalled. "We realized soon that it was going to involve a substantial commitment." An understatement.

The market was still down in 1962, but the amount of wood cut kept going up, coming almost to the billion-board-foot mark, even though 200 mills weren't producing, 80 went out of business, and only 545 operated. This increase was possible because a lot of mills had changed from circular saws to gang saws and were modernizing other aspects of their operations. Thursday Lumber, for example, installed piston-activated "log movers" that fed the logs from the landing to the rollway. Another firm replaced a man with an automated log turner that kicked the log into the right position for the saws.

Low lumber prices and the possibility of pulp mills coming in were not the only problems. The annual management report commented:

> *Mills with no planer are at the mercy of "Planer Row" in Prince George. This type of "one wing" production gets paid for 7/4" and 6/4" which may be planed lightly and 1" and 2" scant boards manufactured from this stock. Thus the planer mill may gain up to one eighth on the lumber talley, but doubtless undersells its competitors who produce a standard finish product. However the number of sawmills producing in the latter manner seem to be dwindling....*

The number was dwindling because by the end of 1963 the image of the future had become clarified. In 1964 pulp mills being built and pulp mills announced were estimated at $1.2 billion, providing over 8,000 jobs and doubling the pulp production in the province.

In addition to the mills already discussed, Greater Peace Forest Products wanted to start a pulp mill in Fort St. John, Peace River Kraft wanted another in the same town, and United Pulp and Price Brothers wanted to build one in Squamish. There were proposals for the Grand Forks-Greenwood area, Crestbrook was going to go ahead and build a mill without a Pulpwood Harvesting Agreement, and there were dozens of other proposals. Some of them were accepted, including Weldwood and Price Brothers' joint venture, Cariboo Pulp and Paper in Quesnel, the Crestbrook Mill and Intercontinental Pulp in Prince George, which was a joint venture of Prince George Pulp and Feldemuhle.

Alexandra Forest Industries, Mackenzie.
Photo courtesy of B.C. Ministry of Forests.

Park Brothers Sawmill at Summit Lake, 1968. Photo courtesy of B.C. Ministry of Forests.

The number of mills operating in 1964 dropped to 450 from the 730 that had been operating in 1955. Firms that owned pulp companies were doing a lot of buying. By running both types of firms, stockholders could improve their control over an operating area and let the decision whether to go for high lumber output or high chip output depend on market prices for lumber and pulp. Firms were paying $10 to $40 per hundred cubic feet for quota, depending on where the timber was, the species involved, the quality—and how much the buyer wanted it.

To further restrict competition, the government brought in a new rule calling for a nonrefundable bidding fee. This meant that if you bid on timber and someone outbid you, you lost hard cash as well as the right to cut the timber.

Northwood's pulp mill was under construction. A thousand people were at work building Prince George Pulp. The cost of living in the north was almost double that on the Coast.

Roy Spurr, at Eagle Lake, went out to the woods where a horse was hauling logs.

"Take him to the barn," Roy said. "Then take good care of him. That's the last logging horse in the north and he's earned his rest." Roy kept the horse until it died of old age.

Prince George Pulp, the first mill approved and the first completed, started production May 1, 1966. When September came and Mackenzie's $6.25 million sawmill went into production to produce sixty million feet a year, Northwood was still sorting out problems with its pulp mill that had officially been running since July. The firm continued to have trouble for years. Weldwood was rebuilding its sawmill that had burned down earlier in the year. And Ray Williston was demonstrating that while the carrot works, it's always helpful to have a stick available.

In 1967 the Forest Products Laboratory in Vancouver estimated that fifty-one percent of a log became finished lumber, fifteen percent became sawdust, eleven percent turned into shavings, and the rest was waste, at most sawmills. Williston had doubled utilization at mills that responded to his offer of more quota in return for close utilization. But some mills hadn't responded. In 1968 the close utilization standards became mandatory in some areas.

One of the reasons mills didn't respond was age. A lot of the mills were old mills, too old to modernize easily. This made combining operations to gain efficiency difficult. Some firms were happy when their mills burned down, clearing the way for a modern replacement and providing insurance money.

When negotiations opened with the IWA in 1967, the union asked to have its contract expire on the same day as the IWA contract on the Coast. The union realized that since many companies had operations throughout the province, it didn't make sense to have part of a firm running while another was on strike. The economic clout of a strike would have less effect. In contrast, if the union could eventually end up with one contract for the entire province, the strike threat would mean a lot more. Getting the same expiration date across the province was the first step.

Company management didn't like the idea.

The union asked for a raise in pay of 50 cents per hour. The companies offered 20 cents. Judge Craig Munroe, the Industrial Inquiry Commissioner, recommended 44 cents.

When a strike vote was taken, only workers at Northwood and in Mackenzie had a majority in favor of striking. On October 4, the men working for Northwood and in Mackenzie walked off their jobs. It was a legal strike.

It was in the best interest of the Northern Interior Lumbermen's

Association to drag out the negotiations. Most of the members had their mills in operation, and the strikes in Mackenzie and at Northwood's Eagle Lake, McGregor, Upper Fraser, and Prince George operations were draining the union's bank account.

The IWA had to drag out negotiations as well. The workers who had voted to accept the contract were not going to walk out in support of "the hotheads," as members called them. And those on strike weren't willing to go back to work, even though the union didn't think they could get a better deal. Stories about the strike were on page one, then on page three, then page eight, then page twenty-four, and finally stopped appearing in the newspaper.

To "help" the strikers, other mills gave their men overtime. Then management said, "Why freeze in the cold? You could be bringing home a paycheck now and retroactive pay hikes later."

To add pressure, management ran an ad on November 6 saying the union had until November 15 to sign the contract. If the IWA didn't sign, pay hikes retroactive to the expiration date of the old contract wouldn't be given. At plants still operating, this ad was not greeted with enthusiasm. The men began putting pressure on the strikers. On the fourteenth the union signed the contract.

The workers at Northwood wanted something that couldn't be put into a contract: a reasonable management attitude. They didn't want the confrontation labor relations Noranda used at many of its mining operations. The workers in Mackenzie wanted something else no contract would provide: a town. Both groups felt that they had been let down by the IWA, and both groups soon signed up with other unions. The acrimonious debate between Northwood and the other members of NILA regarding the handling of the strike led Northwood to leave the association. It still negotiates its own contract with the IWA, although there's rarely any significant difference between the one the company signs and the ones signed by other mills in the north.

Lumber-drying kiln, Prince George, 1968.
Photo courtesy of B.C. Ministry of Forests.

Chapter 20
The End of the Boom

A lot of people expected the boom in the north to last forever, but it ended in 1968. New mills built by companies or planned by consultants used to dealing with the far larger coastal logs ran into trouble. Every time a large-log mill was simply scaled down for small logs, the proudly announced, shiny new sawmill would break down, jam up, cost extra money, and be inefficient. Coastal mills work with different species, like cedar, where there's added profit in "cutting for clears" to get knot-free wood with high value. No one was paying for this grade of spruce, pine, or fir, however. The price of the smaller Interior wood was often far lower than the price of wood from the giant coastal trees. But the profits that could be made were often lower too. So the aim in the north was to process a maximum number of logs per hour. And that was the only aim at a lot of mills.

When the market went down, even Interior-style mills couldn't save firms with poor management or well-run firms that were undercapitalized. Ray Williston told of the result:

> *Everyone doesn't make it the first time. They learn. Nick Van Drimmelen built Netherlands. He got a good Coast consultant to plan a new sawmill and ended up with a first-class problem. Same thing happened to Percy Church. Nick ended up so that RoyNat (the Royal Bank business financing subsidiary) was going to take over. Then they were going to move in on Lloyd Brothers and Clear Lake. Nick had everything tied up, his car, his home.*

In all, five firms in the north were threatened with foreclosure by the Royal Bank. All owed the government stumpage payments. All were in trouble.

> *Gardner of the Royal Bank came into my office. Now I'd let Nick extend his stumpage account into the hundreds of thousands. I told Gardner, "I hope you enjoy owning a second-hand sawmill without any timber."*
>
> *"What do you mean?" he said.*
>
> *I showed him that the Minister had to approve any transfer of timber rights and told him, "I won't transfer a stick!"*
>
> *"You wouldn't do that," he said, then said it again.*
>
> *"Try me."*

"If you ever need fifty cents," Gardner said, "I'll make sure you won't get it."

Williston was within his rights, of course. The Forest Act had a provision that said the Minister had to approve any transfer of timber rights or licences on Crown land. If the Royal Bank had pressed the matter, it could have ended up with five sawmills but no timber. However, the Forest Act of the time didn't include provisions for the Minister to control transfer of shares. Gardner might have been able to gain control through this route if he'd known the Forest Act.

After discussion, Williston and Gardner agreed that the firms in trouble should be placed in receivership so that new management could be brought in, but without bank ownership of the mills. Most of the firms were able to climb back toward or into the black as a result.

We cooled down and it ended up with Ian Bell getting his first major assignment (as a receiver). He's now running Finlay Forest Industries in Mackenzie. They put Gordon Brownridge in at Clear Lake and it worked out well for everyone.

Gardner and I ended up laughing about that meeting in years to come. And, of course, we ended up getting paid every penny owed on stumpage.

That was the year Price backed out of its deal with Weldwood to build Cariboo Pulp, though Price ended up being replaced by Daishowa. But there was good news that year too. The town of Mackenzie got paved streets, a school, an ice rink, and a shopping center. At Fort St. James an $8 million sawmill and plywood plant were announced the day the PGE railway extension to the town was officially opened.

To cope with the boom, the Forest Service staff had expanded each year since 1960 until it finally broke 300 in 1969. The staff was needed because Ray Williston wasn't resting on his laurels. He told the legislature in 1969, "In one SYU in the Interior, this uncommitted balance amounted to more than fifty percent of the allowable cut at close utilization." This amount was allegedly due to improved inventory methods. It was also due to the lack of mills going for the carrot. The government began to require close utilization, while still holding out Timber Sale Harvesting Licences for third band timber as an incentive.

Although few people—other than George Nagle, who was working on his Ph.D. thesis—noticed at the time, this was a major government change. Now the government was saying how logs would be processed, in addition to telling the mills what could be cut, where it could be cut, and when it could be cut. The requirement for barkers and chippers moved government control into the mills.

The five firms that found themselves in trouble in 1968 may have signaled the end of the boom, but the ripples continued well into the 1970s. At the start of the decade construction began on B.C. Forest Products' 500-ton-a-day pulp mill, along with a plywood plant and sawmill. North Central Plywood was started by a group of local loggers: Don Flynn, Jerry Flynn, Gordon Geddes, Howard Lloyd, John Martin and Pat Martin, Gordon Swankey, Dean Shaw, and Chris and Eric Winther. The mill was eventually bought by Northwood, providing a handsome profit for the partners.

But disturbing echoes from 1968 continued to reverberate. After spending more than a million dollars on roads alone, Bulkley Valley Forest Industries had trouble with its $24 million sawmill from the moment it opened. The firm had a Pulpwood Harvesting Agreement that provided rights over 9,400 square miles of spruce, pine, and balsam stretching from Endako to the Coast range, which—considering the value of quota—should have provided an excellent line of credit.

However, the mill was designed for logs of a certain size that were round and didn't taper. The logs it got weren't round, were tapered like all trees from fat butt to narrow top, and were, on the whole, a lot smaller than the mill design could cope with. In addition, "the largest sawmill under one roof in the world" had an inelastic production line: if a jam-up occurred at one point, the entire line would have to stop. The mill lost $5.5 million in the first six months of 1971, then $4.5 million in the second half. After Bulkley Valley had lost a total of $30 million, Northwood picked it up for little more than the firm's debts.

This may sound as though Northwood sat around waiting for firms to get into trouble. It didn't. Northwood had enough financial backing from its owners, Noranda and Mead Paper of Dayton, Ohio, to be able to undertake rescue operations on companies no one else wanted. And Northwood didn't wait for trouble before it invested. In 1971 the two firms that owned Northwood each bought about twenty-nine percent of British Columbia Forest Products, a very healthy firm. In 1970 Northwood bought the first automated faller in the north and changed the way logs were handled throughout the Interior. This automated faller worked like giant pruning shears. It snipped trees at the base while clamps—like a giant pair of pliers—held the tree upright and let the shear operator move it wherever he wanted after the three or four seconds needed to cut through the tree.

The idea caught on. Two firms in Prince George, Muirhead Machinery and Q.M. Industries, soon dominated the mechanized harvesting market in British Columbia. Throughout the north it is now common to see a machine operator listening to rock music through a stereo headset in his air-conditioned cab as he cuts and places a tree a minute. At some operations, chain saw heads are used instead of shears, since they do less damage to the butt of the tree. Watching either type of automated faller at work is enough to make the old-time Coast logger moan softly and head for the nearest bar.

An old-time mill man would be just as shocked. In the 1960s Canadian Car, a division of Hawker Siddley, began producing something called the Chip-N-Saw. Although it had been tested and shown to be effective by a large

Upper Fraser, 1950's. Photo courtesy B.C. Provincial Archives.

coastal firm, the company said, "No thanks, works fine, but forget it." Ernst Forest Products in Quesnel bought one and discovered it was great for Interior lumber. Instead of sawing off one side of a log so it is flat and can be further processed, a Chip-N-Saw literally chips away at the bottom until it's flat, providing, in a single operation, chips that can be sold to pulp mills and a log that can be handled. This idea of chipping away at the log may sound as slow as a beaver in a forest, but the small logs in the Interior of the province literally whiz through, barely pausing. Today the Chip-N-Saw has been improved. A log goes in one end, is electronically scanned to find which face should be the bottom, is flipped into position, and then has up to three sides chipped flat before the log—called a cant once the sides and bottom are square—moves on to the next machine. By 1970 every mill modernization project in the north included a Chip-N-Saw, since John Ernst had demonstrated the machine was effective on small wood.

At the beginning of the decade there were 135 operating sawmills in the Prince George District, with a total of 44 chippers and 52 barkers, plus Chip-N-Saws. The next year some land was taken away from the district to form part of the newly created Cariboo District. In the smaller Prince George District 200 mills sat, not operating. Most of these were either rusting or rusted. Their era had ended. So had the boom.

Chapter 21
Portraits of the Survivors

Adam Zimmerman, Executive Vice-President, Noranda; President, Northwood; Vice-Chairman, MacMillan Bloedel.

Many references to the "forest industry" imply a massive monolith, as though one man or a group of identical men made all the decisions. But once you get away from the concrete canyons of Vancouver and the peer pressure of industry executives who lunch together, you realize that each company has its own identity. One of the standard lines in the industry is "northern companies have more character than those in the south." That's because they're full of characters.

One millowner in the north got a lot of complaints from the workers about his parking lot. Eventually it became an issue, which ended up in the firm's contract with a line something like, "any parking facilities provided must be paved." So the millowner moved a building onto the parking area. No need to pave. The men parked on the street. Then the millowner had the cars towed away for parking too long.

Another millowner in Prince George took a vacation in Hawaii. He got bored after a week, so he began visiting sawmills. One was having problems because the machinery wasn't running properly and wasn't synchronized. He pitched in to help. Instead of coming back after two weeks, he stayed until he had the mill operating properly—just for fun.

The mills in the Northern Interior, Cariboo, and parts of the Southern Interior are usually owned and run by one man, rather than a corporate organization. The typical firm is smaller than its coastal counterpart, so one man can look after the entire operation, at least checking on everything going on. There are exceptions, of course. Canadian Forest Products and British Columbia Forest Products on the Coast each have a strong president, with all operations reflecting the president's philosophy. But in any giant organization, the ideas of the president have to filter through many men before being translated into action. In the north, one-man direct control is typical.

The following four profiles give some idea of the divergence of opinions and attitudes in the north.

NORTHWOOD PULP & TIMBER

Northwood is the most visible firm in the Northern Interior. Its logging operations start west of Smithers and continue east past Burns Lake, then skip to Prince George and continue all the way to the Alberta border. With a 450,000-acre Tree Farm Licence and cutting rights that translate into about two million acres, it's difficult to imagine how large the firm really is. Try 9,600 square miles—just over 2½ percent of British Columbia. You can get in a car in Prince George and drive for hours without leaving Northwood country.

Half the firm is owned by Noranda, which has a reputation for lousy labor relations, and half by Mead Paper of Dayton, Ohio. But Northwood has a distinct personality, reflecting the ideas of one man. Since its beginning, Adam Zimmerman has been in charge.

Zimmerman has been the head of a major B.C. forest company longer than any other chief executive officer. At his office in Toronto, it's obvious he does not suffer fools gladly.

"Look," he says, "you wanted to know how Northwood got started. That'll take some time. We better get going." Adam leans forward and tells his version of the story without the need for a single question.

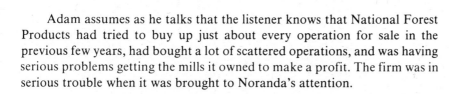
Inside Northwood's headquarters, Prince George.

> Northwood really began when we were brought a rescue operation. National Forest Products had a reserve on timber that became Pulp Harvesting Area #3. Alf Powis, assistant to the president of Noranda at the time, and the treasurer went out and saw the potential. Then I went out and we started to do the deal. David Davenport, Bern Brynlesen- who set the mood and the tone-and I did the deal.

Adam assumes as he talks that the listener knows that National Forest Products had tried to buy up just about every operation for sale in the previous few years, had bought a lot of scattered operations, and was having serious problems getting the mills it owned to make a profit. The firm was in serious trouble when it was brought to Noranda's attention.

> We took on the debts of the company and left the vendors with a residual fifteen percent interest which we eventually bought out for the then undisclosed liabilities of the company. Since I was so involved with the deal I quickly became the person at Noranda who knew.
>
> It took a very tolerant shareholder to let us go ahead. It took ten years before we broke even.
>
> We started out by trying to sort out the people and plants involved. There was Upper Fraser, and an operation at Penticton, Summerland Box, and operations at Osoyoos, Oliver, Sinclair Mills, and others that were derelicts. We were lucky—we had a rich parent, so were able to hang in there.

Llowell Johnson at Northwood's first nursery.

At this point the firm was making lumber and only lumber. Mills were rebuilt, then other mills were closed down since the new plants had the capacity to handle what National Forest Products had run through two, three, or even four obsolete sawmills because of lack of capital for modernization and consolidation.

> *After a couple of years we'd sorted through potential partners for a pulp mill. You might say Mead was the only firm that was willing. Our original idea was that we'd keep the sawmills and participate in pulp only in a minority way. Eventually Mead bought in at fifty-fifty on the whole operation, and it's been perfect.*

In other words, Noranda had wanted to keep the lumber operation for itself, getting in a firm that knew pulp and paper to build a mill and participating only enough to benefit from the chips it provided as they gained in value through processing, rather than just sell the chips outright. Mead, however, wanted to bind its risk by becoming half owners of both the pulp and lumber parts of the business.

> *We thought the pulp mill would solve all our problems. But it was a short-term disaster. It was as large as any single line pulp mill built at the time, but it had no elasticity in the system and some units wouldn't run. We couldn't make the digester work. Noranda's then chairman had insisted on using only things that had been used before and we believed the digester had been used before. It had, but on a smaller scale. Due to scaling up it was bigger than any one like it ever built. We lost $22 million on the pulp mill and sawmills before we got it turned around, and of that I think $12 million to $14 million was on the pulp mill.*

The digester, the giant pressure cooker where chips are "cooked" until the fibers are loosened, is the heart of any pulp mill. If it doesn't work, the mill doesn't work. Northwood's digester was a scaled up version of a small efficient one. You could compare it to a bicycle that has been built exactly like a racer but is twenty feet tall. A lot of problems immediately become apparent—even if you figure out a way to have your feet reach the pedals. Northwood's digester worked as poorly as that big bike.

Some of the mills that produced lumber caused problems too. In some cases they were so scattered that it was difficult to sort them out. Northwood ended up selling its operations in the Southern Interior to Weyerhauser and concentrating on the north.

> *Pulp mills were a waste disposal unit. You have to remember that at the time sawmills were scattered, small, and inefficient. Many were broken down because most individuals were not prepared to invest in them. Most of the individuals who sold out considered us as Santa Claus. And, oh yes, we also bought Giscome, Shelley, Fichtner, Church, and Bulkley Valley. I really have had a significant part in every deal we've done. We haven't been rapacious or attempted to dominate.*
>
> *You know at the time no one really knew what they were getting into. The government was very generous. The*

> *small wood came on very quickly. The government was clever. They never defined a pulp log. And pulp logs became sawlogs. Williston had a lot to play with, but he knew what he had to play with: instinctively.*
>
> *It wasn't planned. We tried to maintain our independence and not get sucked into the Vancouver Club. I think the real thing brought was a state of mind—a miner brings a long term view. He's willing to wait for the end of the rainbow.*

Though the interview was on the record, Zimmerman didn't like being quoted as saying, "We tried to maintain our independence and not get sucked into the Vancouver Club." But that line is the key to Northwood's success. The firm has been, and continues to be, a maverick—which you have to remember originally meant an unbranded steer. Northwood has been of British Columbia, but not part of the Council of Forest Industries. It negotiates its own contract with the unions. While most firms do their best to follow government guidelines that are implied rather than written, Northwood has used the written rules to its own advantage. When the Company applied for a mill licence for expansion of its pulp facility, for instance, the government—under its own regulations—had to say no within thirty days or else the firm would be able to go ahead. A provincial election was called; the government didn't want to be branded as being against economic development during an election; and Northwood announced, just a few days before the election, when the time limit had expired, that it was going ahead. Some members of the Forest Service said unkind things about the company, since a go-ahead on the mill expansion implied the Forest Service would have to assure a chip supply. The rules on advance notice were quickly changed, and chip direction policies of the government had to change to meet the new situation.

Another example occurred in 1981. Many firms wanted to bid on MacMillan Bloedel but knew that the government thought it had a right to approve bidders, even though this "right" wasn't covered by regulation or legislation. Zimmerman talked to the Minister of Forests, who indicated that he didn't like the idea of Noranda going after a substantial interest in the company. Then Noranda went after the shares it wanted, revealing both that the rules didn't exist and that many firms in the industry were frightened of implied government powers that were outside the law.

Adam Zimmerman is respected in the forest industry in British Columbia, but he's definitely not considered "one of the boys." If he were, Northwood could not have been as successful as it has been.

That's part of the story. Doug Little, vice-president of Forest Operations at Northwood, is a history buff who has been with the company since almost the beginning. At his office in one of a handful of major forest company headquarters built of B.C. wood, he listed many of the same firms involved in the beginning of Northwood but added Tulameen Forest Products and Brookmere Logging.

> *These firms all had a common background. Promoters had determined the companies had large amounts of cash, so they took them over and gutted them. Except for a few like Sinclair they were broke or had a poor operating record.*
>
> *People were thinking of a pulp mill in the area, and in December 1963, after Mead came in, we needed some*

samples. We had a lot of trouble finding a chipper. Wood was trucked from Upper Fraser to Kelowna, chipped, and sent to Chilicothe, Ohio.

Buckley Valley Forest Industries, Houston, 1971. Photo courtesy of B.C. Ministry of Forests.

Once a pulp mill site was decided on, and hearings were held for a Pulpwood Harvesting Agreement in May 1964, the newly formed firm bought Fichtner at Newlands, Percy Church at McGregor, Nance at Red Rock, and, in 1966, Eagle Lake Sawmills, which had been the biggest firm in the area since the 1920s. Penny Spruce's tenure came with Eagle Lake: "Before Northwood, Sinclair and Upper Fraser were up against the wall when dealing with Eagle Lake."

This requires some translation. If a small firm is operating in an area that's tucked out of the way, far from a lot of other mills but with one big one nearby, the large firm can refuse to follow the usual industry practice of trading logs that don't fit the small firm's mill—because they're too big, for example—forcing the small mill to haul the logs that don't meet its needs long distances in order to sell to a firm that can use them. One of the simplest examples is a plywood plant. It wants logs to peel. Not all logs growing in an area where it has a licence will be suitable for peeling, but they might be good sawlogs. If the veneer or plywood plant can trade with a neighboring firm, it would make more money than if it had to haul all its sawlogs a long way. Zimmerman was saying that Eagle Lake could buy from its neighboring firms at low prices—just higher than the price in Prince George minus the cost of hauling to Prince George.

Northwood's involvement with one of its largest acquisitions was different from its involvement with those it took over that were too small to make money. Buck River Lumber Company formed Bulkly Valley Pulp and Timber when the idea of Pulp Harvesting Areas became popular. The firm was granted one, then began looking for a source of capital. Consolidated Bathurst and Bowaters Canadian Corporation provided the capital, bought out most of the existing sawmills in the area, and built a new mill at Houston, which has already been described, in part.

The engineer in charge of the project designed the mill to process identical cylinders. This design worked well for some of the spruce that had little taper and didn't vary much in size, but logs aren't often truly round. A lot of things affect tree growth—which side of the hill the tree grows on, whether there are other trees or rocks that block sun and force the tree to

Buckley Valley's new mill, "the largest in one building in the world," 1971. Photo courtesy B.C. Ministry of Forests.

curve toward the light, different soils—and these cause many species to have almost egg-shaped cross-sections. The mill wasn't designed to handle the variations of real trees. The firm became technically bankrupt but was sustained by the parent companies until Northwood bought it in February 1972.

Northwood's solution was to build a new large-log line for oversize logs and a small-log line for undersize logs. That removed many of the causes of the problems. Gradually the original lines were modified so that they could handle trees instead of cylinders.

The statistics of Northwood's operations make a lot of other firms envious. In 1978, before acquiring North Central Plywood, the company had about 2,500 employees and produced over 592 million board feet of lumber per year, plus 710 metric tons of pulp per day. In the words of the Northwood News, that's "enough pulp to make the volume of paper required to print all editions of a magazine for almost half a century" each year.

A lot of people inside and outside the industry feel uneasy about Northwood. One reason is the sheer size of the firm. Another is the company's independence. Some people see the firm as a foreigner, an American (or eastern Canadian) firm that doesn't understand the north and doesn't want to. "When Canfor came in," said one Prince George resident who had been in town before the boom, "they seemed to take over small mills and keep them going. Northwood took everything apart, and put it back together in a different way. It often seems as though they have no heart, just a balance sheet."

In the mid-seventies, Northwood's public stance conveyed the impression that everything was done without emotion, as though by a computer. Even when the company sponsored studies with the Department of Fish & Wildlife, built hiking trails, fought to improve the forestry practices, or initiated other very positive programs, it had a poor public image. Talking with many management people, one received the impression that if anyone had a sense of humor, he left it home.

The company had greatly improved its industrial relations since the late 1960s. Turnover had dropped, and a number of very bright people had joined management. But even in relaxed conversation the management team, with a few notable exceptions, was largely inarticulate.

Toward the end of the decade, however, Northwood began to change. People began to express their concerns more openly. The company began learning how to communicate—with halting speech at first, of course. Northwood still all too often responded to criticism with silence, even when the firm had an excellent story to tell, and at times seemed to have a split personality. But for the first time there was an effort to be understood by "outsiders."

SINCLAR ENTERPRISES

The only lumber wholesaler in the north, Sinclar Enterprises, is related to Lakeland Mills, Apollo Forest Products, Nechako Lumber, L & M Lumber, and other enterprises. The key figures in a variety of businesses with interlocking ownership are Bob Stewart, Ivan Anderson, and the Killy family. They've been involved in the lumber business longer than any other surviving firms in the north, and the many companies they've created are all doing well. In Vancouver board rooms, and in government offices in Victoria, they're considered some of the shrewdest operators in the industry.

Bob and Ivan have agreed to talk about their companies. Bob sits at his desk looking like a displaced fundamentalist minister, tall, thin, and dour. Ivan, whose crew cut and manner give an impression of youth, is relaxed and laughing as he brings in a chair. An interviewer gets the impression that this is a well-practiced act: the nice guy/mean guy pair often used by the police when interrogating prisoners.

> Bob: *In 1942 I left a job as a desk clerk at the Sylvia Court Hotel (in Vancouver) and came north to work for Don McPhee at Sinclair Mills. There were houses for the married couples, and it was the nicest sawmill community in the area. I stayed until 1959, at which time I was assistant manager and Ivan Anderson was manager. When the mill was sold to a group including George Richardson and John Lutton, I went to Oliver Sawmills in Oliver to handle sales. Until then I'd also been involved with Upper Fraser, which Don McPhee had started with Roy Spurr and Edwin R. Safford, Sr., and with Eagle Lake, which McPhee and Spurr bought. At Upper Fraser there were tramways twelve to sixteen feet off the ground. Inside the mill there was a circular carriage with gunshot feed. It was steam powered with two men on the rig. One day the steam valve for the carriage feed stuck and it struck the bumper full force, and when it bounced back the dogger fell between the bumper and the carriage. The man was then caught between the carriage and the bumper when returned, causing his death.*

"Gunshot feed"? "The bumper"? The old mills had ways to kill people that would be laughed out of horror movies as ridiculous. Imagine a giant circular saw—like the one the hero saves the maiden from in old movies. Obviously that's too big to move through the logs like a Skil saw, so the carriage moved instead, taking logs along with it. In addition to the sawyer, who moved the carriage back and forth to cut the proper width of wood on each pass, two other men rode along to keep the log, or several logs at once at some mills, in place. The carriage was steam powered, and there was a bumper at the end of the track to stop the carriage from going through the wall. In the incident Bob spoke of the carriage hurtled down the track, bounced off the bumper, then continued forward again because the steam valve was stuck.

> Ivan: *You have to remember those were different times. We used to pay $6 a thousand feet for logs (delivered right to the mill) in the water. In 1936 I made 30 cents an hour, and had to pay for my board. Back then mills were making up for the money they had lost sawing logs through what was paid in the cookhouse and stable. Charge $1.10 a day room and board in camp and a mill could make a profit. The mills were scattered then. But gradually more and more moved into town and sold their lumber to planers. Before the dry kilns came on the scene, it was nothing to have twenty, twenty-five million feet drying next to a mill.*
>
> *Then, when Lars Strom decided to dry, to put in a kiln, he got less money than people who shipped green lumber earlier that year.*
>
> *Out at Sinclair Mills isolation meant community spirit. There was no road in until 1941 or '42 and it was terrible when they built it. It was about then that the planers started up, Martin Caine, Strom. Northern Planers came later. Martin Caine had a bush mill in town, the Alexander*

> *Lumber Company on Island Cache—that was taken over by Lamb Brothers.*

Question: *But what about your own business?*

Bob: *I came back in 1961 and we started Sinclar Enterprises as a wholesale lumber buyer.*

Ivan: *I was manager at Sinclair Mills and Upper Fraser, and left in May of 1962, and that was the first year I bought lumber. We got a $25,000 loan from the Bank of Commerce. Then it was easier to get credit. A guy could get $2,000-$3,000 and start a mill. Today you need a quarter million minimum, assuming you could get timber. We had a five percent interest in Lakeland Mills for years, and, in 1973, with George Killy, we bought them out.*

A lumber wholesaler contracts with mills for lumber and deals with buyers who want lumber, so the mill can concentrate on making boards and the potential buyer, whether a mattress company looking for wood to be used in box springs or a chain of lumber yards based in Chicago, can negotiate with a few wholesalers rather than many mills. In some cases the deals start with buyers who want a particular mix of lumber. The wholesaler finds who can provide it, who's offering the best price, and who can deliver when the buyer wants it. At other times the mills call to say they have a lot of a particular cut and grade of lumber in their yards, or that they don't have anything scheduled for a week three months ahead. Then the wholesalers try to find buyers. In either case they get a small percentage of the selling price, often five percent.

> Ivan: *Back then you used to build for production. Now you build for recovery. We're rebuilding Lakeland now and recovery will increase at least twenty-five percent.*

In the early 1970s, production was the key to sawmills in the north that were built to process as many logs per day as possible. But as speed increases, the amount of wood that's wasted normally goes up. Slowing down the mill lets a man or machine have an extra moment to decide how each log can best be cut to get, for example, the maximum number of two-by-fours, rather than having added waste. And the faster a lot of standard mill machinery runs, the more likely it is to get out of adjustment.

The alternative is to choose machinery that is designed to get the highest recovery from each log. The new Lakeland mill positions the log within 1/1000 of an inch to get that high recovery and produces twenty-five percent more two-by-fours from the same number of logs the old mill handled.

> Ivan: *In September 1977, we bought out Helco and almost doubled our quota, but the new mill will handle it. But now the timber is committed (in the area). In the future we don't expect much change in the amount we'll have.*

> Bob: *It still, somehow, doesn't seem that we've come that far from Sinclair Mills. We still deal with a lot of the same people. As the only lumber wholesalers around, we bought lumber from Netherlands, Lakeland, Rustad, some small bush mills like Buster Rigler's B. & B. Logging—he's been in business for thirty years—and just about everyone else. About ninety percent of the lumber we buy still goes to the U.S., with most of the rest to the United Kingdom.*

> *You have to remember that over half of British Columbia's lumber output now comes from the Interior, and due to higher recovery (than in the past) there's more every year.*
>
> *There are still some areas that are undercut, north of Fort St. James, and out west. Babine and Houston are now increasing production.*
>
> *It's a lot different in the woods now than it was when we started. We're now hauling over seventy-five miles to Apollo (in Fort St. James).*

Question: *The two of you have been in the north a long time. What do you think of the changes in the industry? What caused them?*

Ivan: *We know of people who made a business of bidding on timber—and of being paid off not to bid or to turn it over to another company. If there is really open competition, there's no competition: if a multinational or a strongly financed company wants it, they'll get it. But there really isn't open competition any more. The Forest Service appraises timber to the nearest sawmill. You have to make a bonus bid on top—hoping the market will go up and you can make a profit. Now sales are divided into interest areas, and you can't bid in someone else's interest area—which is a fine system if no one new wants to get into it. There's a steadily declining number of millowners. Today the economics mean concentration (of ownership) and that makes it very difficult for new operations to start up. Quotas are overcut in many areas so it's hard to allow for new people.*

When Ivan talked of "interest areas" he was talking of something outside of legislation or regulation. The government, in an attempt to halt the rich from wiping out the poor, then gradually expanding to cover the province with three or four companies, has made deals with sawmills and privately told firms that they won't be allowed into certain areas but can expand in others. The regions, or the watersheds, or the Public Sustained Yield Units a firm is already operating in are considered its "interest area." And since this isn't a matter of law, it's possible for one firm's interest area to be highly restricted, while another firm's may be considerably larger. This discrepancy restricts competition. To buy a viable sawmilling operation complete with quota takes several million dollars. And in some areas there simply isn't room for new firms, since firms are already cutting more than they should to perpetuate the forest.

Bob: *Of course there is timber the government wants to sell, especially poplar. We shipped five or six cars of aspen—and lost our shirt on it. There is an opportunity there, however.*

When industry people talk about the depletion of the forest, when they talk about timber in the forest industry at all, it is understood that "what we mean are the species that have a developed market and that we know how to produce at a profit"—and in the north this means essentially spruce, pine, and fir. They are not including poplar, willow, birch, or a lot of other species. The forest industry has been very conservative about exploring markets for these other woods. Bob's comment that Sinclair shipped some aspen places the firm among the most venturesome in the north.

> Ivan: *The biggest change is the pulp mills. I was born here and pulp mills had been promised since the 1920s. There was a railroad track to where the first pulp mill was to go back in 1928, and timber sales were sometimes marked "reserved for pulp mill." But we've come a long way from when there were wooden sidewalks. The pulp mills have tended to make the big companies dominant. Northwood, Weldwood, Canfor, and Eurocan—in a lot of the north, pulp is dominating the forest industry. It started in the early 1960s when there were a lot of amalgamations. If a sawmill was too small, or if it was away from the railway, it couldn't afford to make chips. Chips were all hauled by rail at first. So mills had to be on the rail lines....*

While Bob and Ivan were delighted to talk about 1962 when Cooper-Widman, Ralph S. Plant, Oregon Pacific, Balfour Guthrie, and others were lumber wholesalers in Prince George, their comments on how and why they were able to stay when others moved away was an invitation to generalities like, "If we said we were going to do something, we did it." After the same question was asked several times, Bob said:

> *We took $400 a month in wages. We'd buy a single car (load of lumber), then sell it. We, I believe, were more mill oriented than other buyers, and worked harder to sell. If we earned $500, we spent $450.*

His statement gives a clue to some of the reasons Sinclar has been a success. As discussed later, in the chapter called "The High Rollers," wholesalers often buy from a mill at a low price for delivery in three or six months, hoping they can sell at a better price. Talk to almost any lumber wholesaler and you get the impression that the lumber market is bad, it's going to get worse, and the only reason he hasn't hung himself is that he can't afford a rope. Of course when talking with prospective buyers, wholesalers often sound a lot cheerier about the market. When Bob spoke of being "mill oriented" he meant that he didn't try to grind prices down to squeeze every possible nickel out. Having worked in the mills and with the mills, the firm understood the problems the millowners faced. Instead the emphasis was on sales. Some lumber trading companies still won't accept orders for individual carloads of lumber. Others do their hard bargaining with the mill, instead of with the buyer—even though their pay is a percentage of the selling price. Bob's statement implies that his firm avoided these approaches.

However, Sinclar Enterprises is just one of the firms in which Bob and Ivan are active. Another is Lakeland Mills, which had the most efficient sawmill in the province in 1981. Tom Dilworth, the recently retired manager of Lakeland, told the story in 1978.

Tom started in the industry in 1928, worked for 15 cents an hour in a sawmill in 1933, then was a commercial fisherman in Prince Rupert for a winter before moving to Aleza Lake.

> *I worked there seventeen years at the S.B. Trick Lumber Company. The first winter we logged with horses. Four of us took a contract that paid $3 per thousand board feet skidded to the mill. Due to groceries and horse feed, we ended up in the hole. "I think we'll forget that little debt," Trick said. "Our net was up to $15 a thousand."*

In Aleza Lake, Harry Morrison and Ambrose Trick owned a mill and a planer. Sometimes we had to shut one down to run the other. Even the lumber was different in those days. We made some dimension, some shiplap, cove siding, and pattern 105 or 106 went on every home (built in Canada at the time, it seemed).

Lorne Lyle owned the townsite, the store and a small mill. He was the unofficial mayor and felt responsible for his people. Blackburn and Hasafield had a mill there too. The town of Aleza Lake had about twenty-five families, maybe two hundred people in all. It had its own park, its own community hall and covered skating rink. There was a swimming tower and wharf. People used to come from Prince George to dance. The town was known from Jasper to Rupert because there were four or five good moonshiners there. Moonshine was a dollar a bottle. You could have rye or Scotch—because they flavored it....

I stayed on the East Line until 1951 when I came to Prince George. It was like leaving home, but my daughter would soon be ready for high school. When I arrived, Prince George Planing Mills had the first dry kiln in Prince George.

In 1965 I went to Lakeland. Those were times of change. When automatic trimmers came in we bought one. The machine took the bull-work out of the job. You know, until the Forest Service asked for written contracts, we did it all. There was no need for paper.

It was always a gamble. The price of lumber dropped after the war ended, then came back up. In 1950 it dropped again, then went back up in the late fifties, down in 1965, back up in the late sixties.

Question: *What about Lakeland itself?*

Tom: *Lakeland had a tough time surviving. Ray Williston got us our first timber by Trapping Lake, then in the Carp. He'd promised the company before I came that if the firm would build a small mill and handle the small operators' stuff, he'd make sure we had logs. We got by, and barely got quota. (Remember, this was at the tail end of the period when quota rights were available without buying rights from another company. Most areas had been taken up.) We got it because we had performed. I knew people in the industry. When we put up our first quota (sale) at Trapping Lake, another sawmill operator wanted the timber and threatened to oppose us (at the auction). The next day he apologized. But we still needed the quota. I called everyone in the PSYU and told them we really needed the timber. They agreed not to oppose our application.*

Calling operators in a PSYU and getting them to agree not to bid against Lakeland was illegal. Mention of it is included here because the statute of limitations has run out and, more important, because whether the government or the industry at the time admitted it, it's the way the system worked. And still works. Competition for the timber ended because companies decided not to compete, not because the government restricted competition.

Lakeland is owned, as was mentioned, by Bob Stewart, Ivan Anderson, and George Killy. Today the new mill produces studs, the eight-foot-long two-by-fours that are as alike as potato chips to most people. A lot of mills in the north, which used to produce the studs that were shipped across North America to become the supports in the walls of new homes, have been changing their equipment to diversify. The overseas export market requires lumber of different dimensions. Specialty items, ranging from bridge timbers to wood used to build walk-in coolers, are made in the north too. When Lakeland decided to spend $16 million building a mill that would produce about 120 million board feet of studs per year, other millowners thought it was stupid.

However, building a stud mill that concentrates on a single, identical product meant that the task of introducing new technology to the forest industry was simplified. After each log passes a trio of laser systems that look at width, length, and conformation of the wood, a computer sorts through a hundred thousand ways to cut logs—rather than the millions of possibilities in a random-width board mill. The slabs cut from the log pass through an electronic system that compares each one to grading standards and possible ways to cut, so the computer only has to deal with two-by-fours, instead of everything from one-by-twos to two-by-twelves. And, as Bob Stewart said, "We've had some success with studs. Unless you expect houses to not be built at all, they'll be needed. Sometimes it pays to swim against the tide."

WEST FRASER MILLS

In October 1955, Sam, Bill, and Pete Ketchum each put up about $5,000 and bought Two Mile Planing Company in Quesnel. Today the firm they started is the largest northern forest company that's family owned (with the exception of Canfor, perhaps, but Canadian Forest Products is dominated by its coastal operations). And the Ketchum firm has grown faster than any other firm in the north.

Lakeland Mills, Prince George, 1980. Photo by Brock Gable, courtesy of the Prince George *citizen*.

Back in 1955, there were at least a hundred bush mills and small sawmills near town, and Two Mile Planing was just one of the pack, with thirteen employees, including Sam Ketchum. His brothers had stayed with the family lumber business in Seattle. The Ketchums made enough money to buy fifty-five percent of a mill in Williams Lake, along with a trucking company, sales organization, and tangle of other related companies all owned by the same man: Wright Lumber, Wright Trucking, Wright Forest Products, Swetman Fir & Spruce Sales, Swetman Lumber, and Wm. J. Swetman Ltd.

Within five years the brothers bought the rest of the Swetman/Wright shares, started Bowron Lake Lumber Company, and bought out a number of other small mills. Then they bought more. Then more. In some ways the history of the Ketchums' involvement reads like a list of "Who Was Who": W.A. McLeod; Brownmille Lumber Co.; A.L. Patchett & Sons; Green Lake Lumber; Gardner Building Supplies; Quesnel Supply; building materials stores in Langley, Victoria, Shuswap, and Kamloops; Pacific Inland Resources; a mill owned by Domtar at Dawson Creek; and a mill at Chetwynd.

The pattern of acquisitions was a lot different from Northwood's, however, since the Ketchums had to make each new acquisition show enough of a profit to help pay for the next one. That's a simple-sounding idea, one other firms should have been able to copy. They weren't able to emulate the Ketchums, however, because under the leadership of Sam, the firms, combined under the West Fraser Mills title, had an attitude and personality far different from that of others in the north.

Sam Ketchum, in the words of one of his friends, "combined the ability to be a gentleman in the front room with the knowledge of when to step into the back and knock the stuffing out of you. He delighted in charming a cabinet minister in the morning and grinding a hard bargain in the afternoon."

Sam made the big decisions in his head after talking with his staff and carried what was going on in his hip pocket. West Fraser didn't have a formalized organization. The title "vice-president" didn't appear until 1976. Directors' meetings were held mainly to ratify items required by law and often had no written agenda. One advantage, of course, was that West Fraser didn't pay dividends. Because of the U.S. tax implications for Pete and Bill Ketchum, dividends were usually deferred. Although West Fraser isn't a public company and therefore does not have to publish profit figures, the standard industry estimate, heard from several competitors, is this:

> *In terms of getting back a buck and a half from a 50 cent investment, Sam Ketchum could run rings around MacMillan Bloedel. He put together a logical group of companies so that every acquisition either helped strengthen others or else paved the way for the next.*

Sam Ketchum was far more aggressive than most other millowners. At various times the firm investigated buying (or merging in such a way that the Ketchums would continue to control the new firm) Sierra Pacific, Balco Industries, Integrated Wood Products, Canadian Hydrocarbons, Lignum Limited, Jacobson Brothers, Finlay Forest Industries, Crows Nest Resources, and a number of others.

Then, in the fall of 1977, Sam Ketchum died in a helicopter crash.

Doug Johnston, who took over as president of West Fraser, said, "Almost every major company in the province made an approach one way and another, but we were not interested."

There was no reason to be interested. In August 1978 the firm had on hand close to $20 million in certificates, up from $14 million to $16 million in certificates in June. The firm had no debts.

It continued to expand. It rebuilt Fraser Lake Sawmills; purchased the mill at Chetwynd that was then rebuilt; bought thirteen percent of Abitibi, the largest paper producer in the world; started building a thermo-mechanical pulp mill in Quesnel as a joint venture with Daishowa of Japan; and bought forty percent of Eurocan, the Finnish-owned firm that had a pulp mill and lumber operations in the north. In January 1980, British Columbia Resources Investment Corporation estimated the firm was worth $140 million, excluding any possible profits from the pulp mill then under construction in Quesnel.

This tells only what the company did, not how it was able to do it. In Quesnel, the operating committee members agreed to discuss the company. When they were asked, "What makes West Fraser different from other firms?" almost an hour and a half of discussion yielded three significant points. Doug Floyd, vice-president of production, said, "We do what we think is right and let someone else put a label on it." The line sounds almost banal, but it is one of the keys to West Fraser's success. The same attitude of doing what you think should be done, then justifying your actions later on if need be extends from the planermill foreman to the forester in charge of the greenhouse on top of the sawmill. Time and again people said, "No, you don't ask for approval. If it's in your area (of responsibility), you just do what has to be done and argue afterwards." The informal structure preferred by Sam Ketchum, whether by design or not, has led to a tradition of allowing people all the responsibility they want to take and can handle.

The Quesnel mill has less than five percent turnover in personnel per year. Vacancies are filled by promoting people already with the company, "and it's been that way in well over ninety percent of the instances." The firm boasts that it has one of the best apprenticeship programs in the industry, and it runs the only course for planermen in the industry. Some people at the mill have had their jobs since just after Sam Ketchum started in business. "We treat everybody as our equal," Doug said, "and anyone can talk with any one of us, because we're not different. We just have different jobs. People here, in supervision and the others, look on it as their own company." This attitude is rare in the forest industry.

Asking what the company did for the community caused a lot of squirming at the table. But finally examples of community support began to be mentioned: sending people to Outward Bound, supporting little theater, sending kids to Canadian Forestry Association summer camp, supporting every branch of sports students are involved in—and then examples dried up.

The men in the mill mentioned that each operation hosts an event each year for employees from all West Fraser plants: a curling bonspiel, a fishing derby, a golf tournament, and so on.

"You keep asking about the things we've done. You ask if the community knows about the things we've done. We're not here to keep score. I guess we're inward looking," Russ Clinton, vice-president of forestry, commented. "Is that bad?"

Yes.

West Fraser is nervous about being mentioned in the press. The firm does not have a public relations person and does not want one. And, like every other firm in the industry, West Fraser has some things to regret. For a few years it paid a firm in Seattle owned by the Ketchums a far higher commission to sell some of its lumber than is common in the industry. This commission showed on its books as an expense, probably lowering the stumpage paid to the government. Eventually the government found out and stopped this practice simply by specifying the percentage it would allow as commission.

John Whitmer, Vice-President, Balfour-Guthrie.

When this incident finally became public, however, there was nothing good to weigh against it because the public knew nothing else about the company. The public image of West Fraser was based on this incident and a few other items made public against the firm's will. West Fraser responded by becoming even more nervous about publicity.

When the firm is good, it's very good. And there are two ways to look at it. From one point of view the firm is extremely paternalistic and highly secretive. From another it's too busy doing things it considers important to bother with things it considers "frills."

Whether an outsider considers the attitudes good or bad, the family feeling, the assumption that each individual will do what's right rather than play safe, and its "inward looking" approach are why it has been so successful.

NETHERLANDS OVERSEAS MILLS

John Whitmer doesn't look like the vice-president of a firm that cuts 260 million feet of lumber a year, imports coffee, and exports salmon. He looks like a retired truck driver. He's both.

Sitting in what used to be the old Norman M. Smith Lumber Company office—a firm that became part of Netherlands back in the 1960s—John talked about how Netherlands got rolling.

Netherlands Overseas Mills was started in about 1955 at Lone Butte by Nick Van Drimmelen. It was successful and spread to operations at Chase and Revelstoke, Merritt, 100 Mile House, Likely, McAllister. Then in 1963 it bought its first operation in Prince George, which was Interior Spruce Mills. That was followed by Prince George Planing Mills, Northern Spruce Mills, Norman M. Smith, McBride Timber, and a lot of small independent operations—the Martins, the Bernhardt brothers, the Handfords, and a mill in the Willow area.

Then we started to build on the BCR Industrial Site in 1964—it was probably the first mill in Prince George shipping chips. And we were the first mill to start trucking logs into Prince George. We sold all of our southern operations in 1965-1966 and concentrated everything in the Prince George area.

John speaks slowly, carefully, spelling out names, taking things in careful order. His voice is soft. It demands attention more than a shout would since the sounds of the mill and yard next to the office are almost as loud as he is.

I went to work for Netherlands looking after their trucks in 1958—the year they got their first Kenworths, the first in the Interior. I drove truck as well. A couple years down the road I looked after all their mobile equipment and ran an operation for them for a while. In the spring of 1965 I moved up to Prince George, looked after the same things, plus a lot of their road building and one logging show in the Stony Lake area.

When we went into receivership, I went in as general manager, replacing Rollie Ellison, for the receiver, Ian Bell, and continued until we were bought in June of 1970 by

> *Balfour-Guthrie. I'm still doing the same job, though now I'm also responsible for Polar Forest Industries.*
>
> *The company went into receivership November 23, 1967. The mill started up again January 23, 1968, and has run ever since at the same location.*

This leaves out two things. Why was John chosen to get the mill going again? How did the mill get into trouble?

Although John didn't comment on it, the usual version of the story told by people in the industry is that the manager at the time didn't get along with the receiver, and that John knew more about the operation, and the people who worked for the firm, than anyone else. There's another factor. Big, gentle-sounding John Whitmer gives the impression of being easygoing and relaxed. He may indeed be relaxed. But John was the first northern sawmiller chosen to be chairman of the Council of Forest Industries—the biggest manufacturers' association in British Columbia. This didn't happen by accident. John brings his full attention to whatever he is doing: driving a truck or running a company.

The other question, why the mill failed, is more complex. Comments from others in the industry have included, "the mill was overbuilt, by a coastal engineer," "the mill was ahead of its time and had to wait for the market to catch up," plus a half dozen other hypotheses.

> *I don't know if you can say that the reason for going into receivership was that the mill was overbuilt for the wood that was available, because that was in the olden days of growth and because of the understanding Nick had with Williston. "You shut down all them southern mills, Nick, get rid of 'em. Concentrate your efforts in one area. Go for barking and chipping because the day of full utilization is upon you or just around the corner. And if you're going to be part of this industry, this is the direction you're going to have to take."*
>
> *We were scattered all over, from Revelstoke through Merritt and up through the Interior, but even through those years we were producing 120 million feet a year.*
>
> *When we got into trouble, Nick was making a big approach to the export market, lining up his shipping, his own sales fellows in Europe. Then the European market dropped. This market was poor. We were going through growing pains and very high costs.*
>
> *I think everyone was in the same boat. At that time we did the biggest part of our own logging and all the road building. Then we went into receivership and sold off all the road building and logging equipment, and we went to one hundred percent contract logging. We said, "Our business is this: we will manage the forests and we will saw and that is as far as we will get. Let us leave everybody to do what they do best."*

The idea of selling off the equipment and specialties John was most familiar with and hiring contractors as independents to do the logging goes against a standard rule of management: "any manager will concentrate on that which he knows best and will feel uncomfortable with the rest." John was most familiar with the exact equipment being sold off and had the least knowledge of the mill and forestry aspects the firm kept. This is an example of objectivity rare in any industry.

> Our people spend their time finding better ways of manufacturing lumber and getting the best price out of our products.
> And we make quite a range of products. The fact that Nick built the mill for the export market has proven to be a great help. He might have been early, but his concept was right.

The Netherlands Mill, now one of the oldest mills in the Interior, is really two mills side by side. The large-log mill can produce bridge timbers, wood cut to the dimensions needed everywhere in the world, and longer lengths than are usual. At times the mill has even cut timbers to be shipped to Sweden because their mills weren't able to provide long lengths. The small-log mill is another matter. To take a photo inside, a camera has to be set at 1/8 of a second, and the film has to be able to be processed at 2,400 ASA. If you're not a photographer, that means it's damn dim inside. The mills show their age. "What they need," one competitor said, "is a fortuitous fire."

> Nick's concept has taken us through some very bad years, and we've never had to use the red pencil. We've always had the capability to change cuts. Though only about twenty percent of our products are export, on the average, the mill's flexibility lets us cut bridge timbers— and I think we cut the bridge timbers for everybody in the country because we're really the only mill in the area that can do these kinds of things.... Being diversified was the original concept, and I think it's still the right concept.

Lumber sorting at Netherlands Overseas Mill. Photo courtesy of B.C. Ministry of Forests.

Don Currie, vice-president of lumber marketing for Balfour-Guthrie, expanded on what John said:

> *We make what's wanted by specific markets. Three-inch and four-inch lumber is suitable for remanufacturing in the United Kingdom and north continent. To meet market needs we've organized our mills so we can hedge our bets by selling into markets other than the U.S., and that helps us. We try to avoid dealing with the commodity-type market.*

Instead of competing with other firms that export to the United Kingdom like Northwood, MacMillan Bloedel, Seaboard Lumber, and Eacom, Balfour-Guthrie has been extremely successful at finding the "cracks" and "gaps" in the market—the products people want that other firms don't sell. As a result, some of their customers in Scotland, England, France, and Ireland have used the firm as a source of supply for years.

When it comes to marketing, Balfour-Guthrie and its operations—Netherlands, Polar, Clear Lake, and Hixon—are among the best in the industry. But in other areas it's an uneven firm. At Polar, in Bear Lake, the turnover is extremely high and has been for years. At Netherlands the turnover has been far lower. This is largely because of John Whitmer.

Once John was shopping on a Saturday when some of his workers, drunk as the day after payday, stopped him on the street.

"John! How about a little drink?"

"Well, it looks like you've done plenty of that," John said, "but have you had anything to eat?"

"Nah, we been to busy."

"Then why don't you join me?" He took the group to the best restaurant in town at the time. During the meal the men sobered up—and realized that they were in a nice place to eat. With the boss. Suddenly they became nervous. John helped them relax, paid the bill, and went on his way. He never mentioned the subject again. In fact, when the story was brought up during the interview, he was embarrassed.

But it's a story that's been told and retold at Netherlands. Every man feels he can talk to John.

A sociologist once said that the stories new workers are told tell a lot about a company. At Netherlands they do.

Chapter 22
Another View of the Forest: The Eco-Freaks Meet the Dinosaurs

Trucking in the Prince George area, 1980. Photo by Ric Ernst, courtesy of the Prince George *Citizen*.

The forest is a mirror. What people see in it depends on where they've lived, what they've read.

To a city dweller visiting the forest for the first time, the woods are a wall of green, trees against an out-of-focus background of green. Then more green as one walks into the forest. To most people raised in New York, Toronto, or Montreal, any logging road is an adventure with life at risk. They view the forest as threatening, since it is so unlike "normal" life, or "good," for the same reason. The urbanite is amazed to hear an ant walk across a dead leaf, and a city visitor out with someone who knows the woods quickly finds how little he sees. "Hey, did you see that owl?" "What owl?" "Over there." "Where?" "Oops, he's gone."

Gradually, depending on one's formal schooling, a picture comes into focus. But a lot of different pictures are possible.

A hunter may see a forest as cover, a source of food and water, a still background for movement he hopes to spot. He soon learns to tell where game might be, and what areas can be ignored. There are, for example, fewer deer in a tall, dark forest than in one just ten to fifteen years old. A forester has usually been taught to see trees as a crop and will use emotionally loaded terms like "overmature" and "decadent" for trees others may see as "stately" or "beautiful." A camera with a good close-up lens focuses attention on minutiae like hairy cap moss and lichens, while a fishing rod changes the perspective again. And several of these views of the forest may exist simultaneously in the same person.

It's impossible to see the forest "as it is," because the viewer keeps interfering with the viewing. But while few people see just one picture of the forest, even fewer try to learn to see the forest as many ways as possible. Dr. Alan Chambers of the University of British Columbia phrased it well when he said, "The moment people learn to see from each other's viewpoints, most conflicts are resolved."

One of the problems with learning ways to see the forest is that few people see the forest in person before they view it on TV or read books about it. This establishes images they have to overcome.

A fisherman, on a stream bank when the fish weren't biting, said, "When I first took up fly-fishing I was unhappy with it. It wasn't like in the articles in the outdoor magazines, or the books I read. I even gave it up for a while, until I realized that I couldn't have someone else's experiences, I could have only mine. Then, later on, I realized that just about all those books and articles were selected moments from what it's really like—with things left out like leaky waders, and the time a bat swooped down at dusk and took the fly." Attitudes to the forest change as an individual spends more time in the bush.

Attitudes have also changed over the years. Davy Crockett wrote of one outdoor trip, "I didn't know any body could suffer so much and still not die." He thought the nicest thing about the wilderness was when it was cleared for farming. This isn't the traditional image of Davy Crockett, and it raises additional questions.

Many scholars consider *Walden* the most important book in teaching certain attitudes about the outdoors. The book is a work on nature, but the primary nature studied is human nature. It is not a description of how the author lived in the woods. Thoreau omitted mention of tools he borrowed and the times he dropped by the Emersons' just before dinner to talk and get a free meal. Some of the people who knew him, like Nathaniel Hawthorne, considered him useless. In his lifetime he was considered a failure. Robert Louis Stevenson wrote that "Thoreau's thin, penetrating, big-nosed face, even in a bad woodcut, conveys some hint of the limitations of his mind and character...." Should one react to Stevenson's description by despising Thoreau, and the book? Or should one take the book as it is?

The argument about Thoreau's view of the outdoors, and almost every other view ever expressed, has been going on for years. You'll find a résumé of the changes in attitude in Roderick Nash's *Wilderness and the American Mind*.

Approaching the same problem—where attitudes come from—from a different angle, sociologists present evidence that conservation is an upper middle class movement. The desire to preserve large tracts of wilderness exists largely among educated city dwellers who earn over $25,000 a year. In the United States, and in large Canadian cities, people who are uneducated may fish, but fly-fishing is a pursuit of the educated, or at least those with money. City dwellers do more backpacking than people in small towns.

When one looks at northern British Columbia, however, the situation changes. In the United States, twenty-six percent of the people who live in small towns go fishing each year—far more than the percentage of people who live in cities. In the Northern Interior of British Columbia, the number often exceeds forty percent. In the north, the typical fly-fisherman works in a sawmill. About three-quarters of the families in the north own some form of recreational vehicle: boat, snowmobile, trailer, or four-wheel drive vehicle.

This time spent outdoors is another variable that affects the view of the forest. Northerners see the bush differently from people living in Vancouver and Victoria because they know it better. As a doctor in Vanderhoof said, "If you don't like moose, you'll hate living here." Or, as is common for people throughout the Interior to say, "What else is there to do besides spend time in the outdoors?"

But wherever they live, people get ideas about the outdoors that limit their vision—though they often don't realize it.

Foresters, for example, have often said, "To grow a crop without the idea of harvesting it is foolish. You can no more preserve a forest than a field of wheat. One of our biggest problems is to get the public to realize that trees are a crop, like any other."

Who says trees are a crop? Were the trees that are being harvested in British Columbia planted deliberately? However, the concept has been with us for a long time. It can be traced back to, among other things, the destruction of the Danish fleet in the days when Napoleon was ruling France. During the Napoleonic Wars, England found that it had cut down too many of its forests and was unable to supply its ships with masts. So had France. The English got their masts, spars, and timbers from Norway and the New

World and tried to keep France from getting the wood it needed to repair its ships. It's a lot easier to defeat a rotten ship than a sound one. When the Danes said they were going to allow trade of nonwar materials, including wood, England sent Nelson to argue. And in the bloodiest battle of the war the English beat the Danes and made sure that French ships would continue to rot. One result was a flood of books, laws, and forestry schools talking of the need to grow trees. Today this attitude is still popular. The United Nations considers forestry a form of agriculture.

Attitudes of the past continue to intrude on the present. The government's attitude toward the forest started in 1066, when William found that conquering England took a lot of money. So he enacted new laws that said, in effect, "the forest belongs to the King and if you want anything out of it, you'll have to pay me." British Columbia adopted this English attitude that the forest belongs to the Crown, as distinct from the public. And it adopted the belief that the forest existed to provide revenue to the Crown, not necessarily to individuals.

To look at public attitudes, there's no need to go back so far. In 1913 Joe Knowles stripped himself naked and went into the Maine woods, each day leaving notes written on birchbark with berry juice telling how he survived. These notes were printed in newspapers across North America. When he finally came out of the woods, dressed in a bearskin, "survival" had changed from getting through Thursday to something manly, romantic, and exciting.

The next year Joe did the same thing again. But the day the United States declared war on Germany, he came out of the woods to enlist. How did he know the United States had declared war? Allegedly, he was really sitting in a cabin eating steaks and writing his stories on birchbark. People forgot the exposé, because the legend was better.

Other people got some of their ideas from Aldo Leopold. In *Sandy County Almanac,* he popularized the use of the word "ecology." But many of the readers forgot—or never knew—that Leopold was the founder of the science of game management.

Leopold's "land ethic" has been distorted by others, but originally he wrote:

> *Conservation is a state of harmony between man and land. Despite nearly a century of propaganda, conservation still proceeds at a snail's pace; progress still consists largely of letterhead pieties and convention oratory....*
>
> *The "key-log" which must be moved to release the evolutionary process for an ethic is simply this: quit thinking about decent land use as solely an economic problem. Examine each question in terms of what is ethically and esthetically right, as well as what is economically expedient. A thing is right when it tends to preserve the integrity, stability, and beauty of a biotic community. It is wrong when it tends otherwise.*
>
> *It of course goes without saying that economic feasibility limits the tether of what can or can not be done for land. It always has and always will. The fallacy the economic determinists have tied around our neck, and which we now need cast off, is the belief that economics determines **all** land use. This is simply not true. An innumerable host of actions and attitudes, comprising perhaps the bulk of all land relations is determined by the land-users' tastes and predilections, rather than by his purse....*

Some people read Leopold's land ethic, blended it with nostalgia, the reverence for all forms of life popularized by Albert Schweitzer, and a desire to escape from civilization as it is all too often today, and ended up with "a reverence for wood."

A final example of the effect of books: in the early 1970s at a Lands Branch office in the north, a visitor was talking about rules of access when a young couple came in, clutching a copy of Bradford Angier's *How to Live in the Woods on Pennies a Day.*

"Well, we're here!" the couple said. "Where's the free land?"

The officer turned to the man he'd been talking to. "This happens literally hundreds of times each summer. Now do you know why I drink?"

Of course, there's no need to get ideas about the outdoors from books to have a restricted vision of the forest. For decades one of the two main tasks of the Forest Service was to help forest companies get land they wanted. The other, related task was to map the forest. People in the organization at the time usually didn't deal with the public. Instead they dealt only with senior civil servants and the industry, plus a politician now and then. So when, in the early 1970s, the public—or at least a vocal section of the public—demanded the right to tell the Forest Service what to do, the members were dumbfounded. One senior official said, "Our aim is to do our job quietly, and frankly to be left alone."

The forest industry was as ill-prepared.

But they shouldn't have been. The idea of going into the outdoors for fun was new in North America when the U.S. Civil War taught thousands of men how to camp. The first book on the subject to become popular in North America, Murray's *Adventures in the Wilderness,* had a ready audience of war veterans when it was published in 1869. And the book provided the "practical" excuses needed to popularize a hobby, including the claim that the North Woods (then located in Maine) would cure tuberculosis and melancholia. Then, after the Spanish-American War, books like Stewart Edward White's *Camp and Trail,* magazines like *Forest & Stream,* and the introduction of specialized, less bulky, more comfortable camping gear attracted another generation. After World War I, outdoor magazines flourished. After World

Anyone in the north can have a lake to himself...

War II, the sale of sporting goods experienced a boom that didn't really begin to level off until 1980. In 1959 180,000 freshwater fishing licences were sold in British Columbia. By 1973 licence sales had gone up to 426,729.

As numbers increased, so did the number of conflicts between Chryslers and Kenworths on logging roads. Each new logging road that was built provided access to new areas for hunters and fishermen. People pressure became a problem. In the north, away from the major population centers in the province, some lakes were fished out in a single season. Further south one lake near Hope was almost denuded of fish in a month. Popular roads looked like multi-mile trash heaps during hunting season. More people were using the forest. And that meant more people were seeing what the forest companies were doing.

The pressure was increased by the growing environmental awareness in the United States that, carried to Canada in magazines and books, gave Canadians another way to look at their forests. The U.S. reaction to the Viet-Nam War also spilled north. And part of that reaction was similar to what had happened during the Depression, just after World War I, and in the years following the Napoleonic wars. People wanted to get away from an all-too-grim reality, and so they sought peace in nature. Again nature was used as a mirror. People found it pleasing and wanted to keep it as it was, preserved like a photograph.

Near Hope, a logging company put up a gate to keep people out. It was ripped down a week later. The company replaced it with a steel fence and gate. Two days later both were gone. The company put up a steel reinforced concrete pillar with a steel gate forged from railway tracks. It was blown up.

North of Prince George, a firm put signs on a narrow road saying that only logging trucks were allowed. Someone tumbled a section of mountainside onto the road.

On Vancouver Island the RCMP had to be called out to prevent open warfare because of forest companies' "keep out" policies.

Forest companies had always thought of the treed area of the province as theirs and, all too often, as theirs alone. In Tweedsmuir Park, for example, Crown Zellerbach logged right to the edge of a major salmon river—hurting the fish habitat—and put signs on its roads reading, "Private Road, Keep Out."

In December 1972, Dr. Ian McTaggart-Cowan and Roderick Haig-Brown shared a full page in the *Vancouver Sun* pleading for a view of the forest that was wider than the usual "crop" view. In the same paper three months later Dr. Peter Pearse said that hunters were getting shortchanged for the $40 million they spend in British Columbia each year. Every lumber industry convention from the Truck Loggers to the Council of Forest Industries included one speech sure to grab headlines like, "Forest Industries Should Declare War on Wilderness Weirdos." Here is one example, from the January 10, 1973, *Vancouver Sun:*

LOGGER TAKES AXE TO ECO-FREAKS

> "Eco-freaks" should face a public inquiry on their allegations that B.C.'s natural resources are being mishandled by the province's forest industry, Truck Logger's Association's Viv Williams said today.
>
> He denounced "wild-eyed allegations" against the industry by "urban environmental activists who never venture far from their Volkswagen ('busses'....)

Literally thousands of newspaper articles came out during this period demanding, defending, pleading, or commenting on one narrow view of the forest or another.

What is often forgotten about this period of conflict is that although the battle to change attitudes was fought by people who wanted to *use* the wilderness, their support came from people who felt that the *idea* of wilderness was important.

Every political party had environmental resolutions as a major part of each convention. Some were overdue. Others were stupid.

However, elected officials only have to avoid going directly against party policy. They can pick and choose the resolutions they act upon. As a result, actual legislation passed in British Columbia during the height of the fight, 1971-74, was far less than in the States during the same period. This was because the government didn't face the same pressures.

One reason was the sheer size of British Columbia compared with the population that used the outdoors. There were more backpackers living in California in the early 1970s than there were people in British Columbia—and the Californians had far fewer places to go, of course.

The other reason, more important in the long run, was the attitude of a handful of people in the Forest Service. The most important of these was Bill Young, District Forester in Prince George at the time, who became Chief Forester for the province a few years later. Bill and Roger Goodlad, who headed Fish & Wildlife in the Prince George area, pioneered a concept called "the folio system of integrated management use" and tried it out in Prince George. It was different from the older idea called "multiple use," which basically said that "the forest has room for the industry and hunters, fishermen and some others," but didn't define how this would work. In the "folio" system, the foresters put an overlay on a map showing the types of forest in an area and how good they were. Then Fish & Wildlife put on another overlay showing critical wildlife habitats, areas where there were caribou, and so on. Then the Lands Branch put on an overlay showing where people wanted lakefront lots and other developments. Agriculture put on another overlay. The result was a folio for each area showing precisely where the areas of conflict were—and where they weren't. The folio usually revealed that in about ninety percent of the area, problems didn't exist. Representatives from the various government departments would agree on this, and then planning to maximize use of the single-use land could begin. The remaining ten percent was the subject of discussion, argument, and compromise. Usually a reasonable solution—in the opinion of all the special interest groups concerned—could be worked out. Conflict was minimized.

Another project started in the north, under Bill's direction, was dispersed backroads campsites. People in the outdoors tend to herd together. And they

...with trout. Photos: Kathy Bernhohn

tend to be slobs unless they're reminded to put their trash away—in a convenient bright green trash barrel. But at many of the popular lakes in the north the beaches were strewn with trash, and people began to complain to the Forest Service. After all, the Forest Service was responsible for the forest, wasn't it?

At Chief Lake, near Prince George, the Forest Service put up a few trash barrels and cleared a small picnic site. People began to use the site—and to put trash in the barrels. In 1970 Harry Marshall of the Service did a study that showed this approach would help funnel visitors in the forests and would help contain the trash problem. He also found that if each site had a picnic table, the percentage of recreationalists who used the sites would go up, further containing the litter problem. The program began the following year with a handful of sites in each region. By April 1975, there were over 780 in the province. The next year there were 850. Now there are well over a thousand. Each one, whether it's a single site or one with sixty tables, fire pits, and toilets, is designed to handle and react to local pressure, rather than to attract people to an area. The program made people feel that they could get to places enjoyable to visit in British Columbia's forests and that they were welcome. Thus, a lot of antagonism was eliminated.

While Bob Williams was Minister of Forests, he ruled that all roads in the forest under Forest Service control, or paid for by the Forest Service, should be open to the public whenever possible. This ruling helped a lot too.

The forest had been mismanaged. In addition to being unable to actually get a sustained yield forest going in British Columbia, the Forest Service had been—as discussed—for many years a government tool to aid industry and little else. In 1972 there were 6.8 million acres that were officially provincial parks, down from 8.4 million in 1959. Most of this land was in three parks: Tweedsmuir, Wells Gray, and Hamber. But in 1972 Tweedsmuir, the largest park in the province, had only one road through it. The only trails were those built for the extraction of timber, those developed before the area became a park, and a few very rough trails—often marked by sticks leaning against trees—developed by the Sierra Club.

That year B.C. provincial parks had over eight million visitor days. Because of lack of funding for park development, most of the visitors stayed at campgrounds within sound and sight of major highways. The NDP government increased the amount of park acreage by twenty-five percent to 8.5 million acres, just ahead of 1959 figures. However, the land they added was often a three weeks' drive past the end of the earth. Only a few people visited the Mt. Edziza, Tatlatui, and Kadacha wilderness parks between 1973 and 1976.

Access is important. The result of adding parks that were just marks on the map actually increased the demand for parks. The new parks demonstrated that parks could be added. Now people wanted them added within easy visiting distance.

Unfortunately, the groups lobbying for more parks didn't seem to be aware of a lesson the government had painfully learned—that people don't use parks, they use ribbons within the parks. The number of people who have backpacked in Manning Provincial Park, which is within an easy drive of Vancouver, is far less than one percent of the people who have visited the park. Most visitors expect sani-stations and improved sites with outhouses and running water. The possibility of getting away from civilization is often more important than actually getting away from it.

This phenomenon doesn't apply only to parks. The moment you have to leave pavement for a dirt road, usage drops in half. Leave a developed road for a trail, and the number drops in half again. Go a quarter mile off the trail, and usage drops to less than one percent of outdoor recreationalists. This decrease is true for hunters and berry pickers, fishermen and backpackers. There are even a lot of campers who camp to meet other campers rather than to get away from people. A visit to any campsite near Kootenay Lake or in Manning Park will confirm this observation.

The need for developed, accessible sites near population centers still exists. And the need for wilderness that's supported as an idea, though it may never be visited by the supporters, has never faded. In 1981 areas of conflict included the Cascade Wilderness proposal northwest of Manning Park, the Stein and Skagit river valleys, the Stikine River, the southern half of Moresby Island in the Queen Charlottes, a variety of areas in the Kootenays, the North Thompson River watershed, and the Southern Chilcotin plateau. The differences between the situation in 1981 and the one that existed in 1973 are that environmental concerns have become respectable and that none of the conflicts are in the north.

Although some politicians and forest companies have said that the environmental wave has passed, what has really happened is that ideas once considered controversial, such as concern about adequate recreational resources, wildlife conservation, and the need for better forestry (farming) practices, have become familiar to a large segment of the population. They are now "respectable."

The reason for the small number of conflicts in the north, at least in early 1981, is that many of the conflicts have been resolved, one way or the other. Perhaps the place that best shows this is Smithers.

Smithers is a strange town. It has more university graduates than any other town of its size in the province. It has a doctor who is a licenced guide and was head of the local Rod and Gun club. The Information and Education Officer at the Fish & Wildlife office—the publicity man—refused to be interviewed about the outdoors for a radio program sponsored by the forest industry in 1977. There's an environmental foundation twelve miles away at Telkwa and a lot of people of pioneer stock who think their town is being ruined by "progress." Hudson's Bay Mountain, which looms over the town, is one of the two large tracts of land in B.C. set aside by the Forest Service so that motorcyclists and backpackers, snowmobilers and cross-country skiers can enjoy themselves without getting in each other's way.

Considering the attitudes of the "Smithereens," as the residents are sometimes called, one of the most courageous acts Bob Williams performed while he was Minister of Forests was to visit the town and talk in favor of development.

In the mid-1970s, Smithers had a reputation as the town with more environmental activists, and public support for them, than anywhere else in British Columbia. All you had to do was say "Smithers" to people in the forest industry in the north, and they'd shudder at the thought that the ideas popular in the town—like the concept that the townspeople should have the final say on what happened to the forest—might spread.

Smithers deserved its reputation. As a result of public pressure, a large area nearby was set aside as British Columbia's first Integrated Management Unit. Only experimental logging was allowed. The Fish & Wildlife Branch set up a committee with representatives from various groups to thrash out wildlife management problems and develop a management plan. The Forest Service set up a similar group. Both committees have taught the participants that differing views of the forest can be valid. Committee members have all learned new ways to look at the wildlands. As understanding increased, conflicts decreased. Today Smithers is a quiet town, because people involved in the forest have learned multiple ways to look at the woods

Chapter 23
The Saga of the NDP

Dunkley Lumber, May 1971. Photo courtesy of B.C. Ministry of Forests.

Ultimately, the issue may be one of ideology vesus philosophy:
Ideology: *False ideas held by your opponents; a rationalization of their interest positions.*
Philosophy: *What you believe.*
—Economics in Plain English
by Leonard Silk

When the New Democratic Party was elected as the provincial government on August 30, 1972, the forest industry felt threatened. Socialism meant change. To an industry built on vague acts and "Ministerial discretion," the election brought the realization that the government could wipe out any company it wanted.

Bob Williams, the new Minister of Forests, wasn't reassuring on election night when he said, "We've been given a mandate and we're gonna clean house." The industry members who delved into Williams's background found he had written in *Canadian Dimension* the previous November:

> Half of B.C.'s economy is based on the forest industry but the revenue directly obtained from the industry in royalties and logging taxes totals only 7 percent of the provincial budget. Of that 7 percent—$70 million—30 percent goes toward operating the Forest Service. That leaves $40 million as net return for the treasury. Just how small the public's return is can be seen by the fact that liquor sales in B.C. net the provincial government $60 million....

The idea that the government could get more money from the forest was as old as the idea of a governmental forest policy. Although Williams complained of penny-an-acre Tree Farm Licences, artificially low prices for chips paid by pulp mills, a lack of competition for wood, and regional monopolies, he was simply doing the usual thing out-of-government parties had done for years. But part of the NDP's official party policy was to "establish a forestry commission which would...embark upon the exploitation of all forest resources through Crown-owned Corporations...."

The NDP also wanted profit and loss statements published by every forest company and a complete halt to logging in any potential wilderness areas.

No one knew what the new government would do. Premier W.A.C. Bennett had kept the NDP at bay for twenty-two years by warning that "the Socialist Hordes are at the Gates!" Now the threat had become a reality and the forest industry didn't know how to handle the situation.

Neither did the NDP.

If Williams had followed party policy and taken back the forest, the industry would have found it sensible to spend as much as necessary to defeat the government and bring back quota, tenures, and noncompetitive bidding. If Williams had done nothing, the industry—out of fear of what he might do—would have tried to get him. It was a no-win situation.

The problem of being British Columbia's first social democratic (or democratic socialist) government split the caucus. Some NDP Members of the Legislative Assembly believed in "the inevitability of gradualism" that was the basis of Fabian socialism. Others thought the election of the NDP was a fluke and wanted to go "four years full bore." "We wanted to change the consciousness of the province so that British Columbia could never go back to W.A.C. Bennett-style rule," one MLA said. For about the first two-thirds of the NDP's term of office, the "four years full bore" philosophy ruled. It ruled secure in its righteousness, sure that though people might be displeased at the moment, "when they see the good results, they'll understand."

But it was the NDP that didn't understand. Canada has the same attitude toward corporations the United States did in the late 1890s. Canada has the weakest competitions acts in the western world. The attitude of "what's good for General Motors is good for the country" is endorsed by Canadian Pacific, Noranda, retail shop owners, and the Federal Government. There is an almost religious belief in the classic economic view that the corporation and the state have identical goals.

However, Wassily Leontif, the Harvard professor who received the Nobel Prize in economics in 1973, wrote in *Capitalism: The Moving Target:*

> *References to the moral virtues of self-reliance and self-help, to the efficacy of private enterprise...leave no doubt that at the bottom of all this lies the profound belief that most social and economic problems would be solved or simply vanish if I.T.T. and a small homesteader in the Ozarks...were simply encouraged to defend and promote their separate interests by any and all means available to each one of them. Such a belief can be based only on a profound misunderstanding of the system in which all of us live and the forces that keep it moving....*
>
> *The pursuit of private economic gains is certainly the mighty power source that propels the American economy... but to keep it on a chosen course we have to use a rudder. The steering apparatus consists of taxes, subsidies, anti-pollution regulations and other measures of government economic policies....*

This may sound like the type of philosophical discussion one has late at night after the second case of beer. But this question was at the heart of the fight between the NDP and business. It can be phrased several ways, but it basically boils down to an argument over what the corporation owes society and what the government owes business.

Here in Canada, the subject is not discussed in polite company. In the United States it is. David Rockefeller, head of the Chase Manhattan Bank, also wrote in *Capitalism: The Moving Target:*

...corporations must develop more effective tools for measuring the social as well as the economic costs and benefits of their actions.... It is vital that social accountability becomes an integrate part of corporate conduct, rather than a philanthropic add-on. Only in that way will the economic development of the private sector move forward within an acceptable framework of public purpose. Only in that way will corporations assure the healthy social climate vital to their own future prosperity.

In British Columbia, in 1972, such ideas were unheard of. Bob Williams never discussed with the forest industry what the companies owed the government or the province. The forest industry never discussed with the Minister what the government owed the companies. Instead, the two sides talked *at* each other.

The NDP's value system differed from those of previous governments. The forest industry's standards of morality differed from those of the NDP. As a result, it was as though Bob Williams was speaking to the industry in Hebrew and the industry was replying in Gaelic. The less the two sides understood each other, the louder they spoke, like tourists trying to be understood in a foreign country.

This may make the NDP years sound as though they were a dialogue, without understanding, between industry and government. To a great extent, both the NDP and the industry thought it was—until the unions demonstrated the error of their ways.

Let us go back to yesteryear. The curtain is going up on a drama decided before the actors take their places on the stage. The NDP and the industry take their pre-ordained roles. Richard Nixon is running for re-election in the United States. British sex scandals fill newspapers. The average house costs $24,000. Frying chicken is 45 cents a pound, and there's talk of a consumer revolt against high food prices.

The NDP has just been elected, and lumber prices are at record highs.

At the beginning of November, Cariboo Pulp and Paper goes into production at Quesnel. Later in the month Dave Barrett tells a labor meeting, "Don't think heaven has arrived" just because the NDP has been elected.

On December 7, Pearl Harbor day, newspapers announce that the first export of lumber in Japanese dimensions from northern mills will take place early in the new year. B.C. Forest Products' new 500-ton-per-day pulp mill in Mackenzie makes the papers just before Harry Truman and Lester Pearson die.

And in Victoria, Bob Williams is trying to find out what's going on and how it can be handled.

In 1978, speaking of the Forest Service and the civil service in general, he recalled:

We inherited a 1930s-1940s style administration. The people in industry who tried to "help" were in it for themselves. (When you talked to industry representatives,) you always felt you were getting a biased story from them, but that's to be expected. They were playing the game. The industry did vary. It wasn't completely monolithic. Some medium-scale operators—especially the unintegrated sawmillers—gave good input.

It was only natural that everyone would try to pass the biases of special interest group thinking on to the new government. Governments are like

Irishmen as viewed by Jonathan Swift. "Much can be made of an Irishman—if caught young," Swift wrote. The B.C. Independent Loggers' Association, the Council of Forest Industries, the IWA, and environmental groups all knew it would be far easier to implant an attitude where none has existed than to try to change one later on. So everyone rushed to Victoria to give input to the government.

Looking out the window of his book-jammed study, Williams recalled his first impression on becoming the Minister of Forests:

> *The Forest Service left a great deal to be desired. Nobody had a grip on the economics of the industry. There wasn't a single professional economist on staff. Only the head of the appraisal section realized this. He'd taught himself and was the most knowledgeable person around. For example, I asked the deputy minister, "How much has quota traded for in previous years?" to try to find out how much we were undervaluing timber. He replied, "I have no idea."*
>
> *It was an inadequate bureaucracy with modest talent. The industry had scads of talent, but we—meaning me—trying to administer the forests were working in the dark.*
>
> *The problem was that I had a broad brush understanding of the forest industry, but no one to follow through. The ability to take an idea and turn it into a reality just wasn't there, so for a while nothing got done.*
>
> *The industry and the Forest Service were two solitudes. There was no cross fertilization, no crossing over.*

It was a time of waiting. To forestall governmental changes, Gordon Draesaeke, president of the Council of Forest Industries, claimed the industry was already overtaxed. He told the press less than two months after the NDP was elected:

> *During 1969, the previous government, largely as a result of NDP pressure, moved to a sharply higher stumpage and tax policy, with the result becoming obvious both in 1970 and 1971. Return on investment dropped to 2.3 percent in those two years, as a result of the revaluation of the Canadian dollar, poor markets and greatly increased taxation.*

Draesaeke didn't say which were the major causes of the drop in profits. There was no mention that some of the companies in the industry might have been poorly managed. Instead, he concluded with words echoing the testimony before the 1910 Royal Commission on Forestry, words still used whenever the industry is displeased:

> *...the new government can not continue to appropriate such a high share of the available money without serious damage to the industry. By damage, I mean stagnation, less employment—and possible closures.*

A month later, on January 19, the Prince George *Citizen* reported, "Prince George manufacturers share in huge price windfall." In the first two months of the year 60.8 million cubic feet of wood were harvested in the Prince George District, twenty-one percent more than had been cut in the first two months of 1972, as a result of a sharp rise in prices.

In March 1973, the government announced that it was buying the Ocean Falls Corporation, including the town, from Crown Zellerbach. According to the NDP, Crown had the entire town up for sale, was going to close down the mill, and everyone in town would be forced to move out, swelling the ranks of the unemployed. According to the industry version, the government was interfering with free enterprise and violating common sense. It was said that Crown Zellerbach had been forced to take a low price and that the mill would be a drag on government finances.

Williams responded to the industry's upset over Ocean Falls by announcing at the beginning of April that the government was buying seventy-nine percent of Canadian Cellulose Ltd. Celanese Corporation of New York had wanted to sell the firm's Castlegar pulp mill to Weyerhauser, but the government didn't approve the transfer of the Tree Farm Licences involved. Her Majesty's Loyal Opposition said the firm's two Prince Rupert pulp mills had been losing money since 1966 and ran ads headlined, "STATE CONTROL."

Reminiscing about this period, Bob Williams said:

> *The attitude of the companies toward accepting reasonable change made intervention through equity ownership—Kootenay Forest Products, CanCel and Plateau Mills—the easiest way to go. It was clear early on that the idea was to play the game the way they did: pick up a company when available and redirect the timber (combining operations for more efficiency). That's how we played it. I believe in a pluralistic state and their attitude made it the only way to go.*
>
> *I felt I had a job to do, so the screams didn't bother me. What really gets one when you get in these so-called positions of power is the contempt for the process. They didn't consider this kind of power legitimate. The idea that people have the authority to make changes based on the ballot—based on the voters—isn't accepted at all.*

Williams was partially right. Some companies mouthed the words "multiple use" but saw their logging areas as fiefdoms. Other companies felt—and still feel—that the province's timber supply would dry up and so were out to "cut and get out," making money at the moment and forgetting about the future. A lot of companies felt—and still feel—that the government owes the industry a profit in return for providing jobs. Without exception there wasn't a single firm in the north that looked on the voters as having the right to say what should be done with the forest. The forests were theirs—meaning the forest companies'—not the public's, held in trust by the Crown.

But some of the southern firms were aware of the power of the voters. "Everything we can or cannot do depends on the voters," an executive at MacMillan Bloedel said. The firm's glossy newspaper supplements, "A Walk in the Forest," were a direct appeal to the voters.

The price of lumber began to drop in March 1973.

Slipping lumber prices hurt confidence. And the NDP's environmentally oriented logging guidelines, introduced that spring, scared the industry. The guidelines were designed to eliminate logging damage to the environment, but they raised costs as profits dropped. The industry felt the government was "out to get them."

If the guidelines had been introduced in isolation, the situation might have been different. But "four years full bore" meant a rushed introduction of the Land Commission Act designed to save farmland. The original version had an expropriation feature that upset many people, including a lot who were pro-NDP. The government reacted to criticism by pointing out the purity of their motives and by confusing an attack on this section that said the government could take whatever land it wanted with an attack on the idea of saving farmland. Suddenly the government distrusted all criticism. The public subjected the NDP and what it did to a scrutiny usually reserved for *Playboy* centerfolds. The Opposition poured gasoline on the flames by doing things like announcing, falsely, that the NDP was forming a secret police. Soon the forest industry would have distrusted the government even if it had abolished all taxes and given each firm money.

When the government purchased Plateau Mills near Vanderhoof, for example, *B.C. Lumberman* reported the buy-out in its June issue in a manner that more than somewhat lacked objectivity:

> *...Communist regimes gained control of the economy simply by seizing capitalistic enterprises. Socialists, who stop short of such tactics have demonstrated that they can achieve the same end in a more gentile (sic) fashion using corporate and personal tax revenue to take over the province's (someday perhaps the nation's)* private enterprise....*
>
> *This is not a partisan and politician (sic) outpouring. Maybe it is right, maybe it is wrong. On the one hand there is the Utopian idea that wealth (through the government)* will be shared by all, on the other the prospect that individual initiative will be snuffed out at the possible expense of an economy that rewards free-thinking, reward seeking innovators....*

Despite the screams some good ideas were quietly adopted by both sides.

The "folio" idea developed by Bill Young and Roger Goodlad was extended to cover the province so that resource conflict areas could be identified and handled. Although the idea didn't work perfectly in all areas, it was a step toward integrated resource planning and was accomplished with few complaints.

The industry demonstrated its faith in the future in the most meaningful way a business can: with cash. In August 1973 Ivor Killy sold Ferguson Lake Sawmills to Balfour-Guthrie. Jacobson Brothers in Quesnel and some other firms scattered throughout the north continued to invest.

But in October there was a critical shortage of railcars. Both the BCR and CN had ignored rising production in the forest industry and past shortages while seeking short-term gains. The lack of cars restricted the amount of lumber that could be shipped. This restriction caused prices to stay higher than they might have been. But it also caused mill yards to fill with lumber awaiting shipment.

Then, later in the fall, the government announced new stumpage rates.

The public has never understood why the industry raised such a fuss over the new rates. Basically the government decided to admit there were more ways to make money from a tree than just turning it into lumber. When Williston had introduced chipping and pulp mills to the north, the formula on which government revenues were based was not changed, so chips had been left out. Now they were included, although at a low rate. Mills had been improved over the years, getting more lumber from each log, but the formula used to assess payment to the Crown hadn't changed. Now the higher recovery, by tree species, size, and quality, was incorporated.

---

* Parenthetical remark is part of original article.

Van Scoffield, manager of the Council of Forest Industries' northern office, said at the time, "We accept the principle involved, but disagree on the factors in the formula." In other words, the hike in taxes hurt. The Truck Loggers' Association said the new rates were "devoid of economic realism." But it was Harry Gairns, of the Industrial Forestry Service in Prince George, who really understood what the government was after. "The professed aim of the B.C. government," he said, "is to increase the tax until the manufacturers have just enough profit to keep going."

That's true of all governments. Before World War I Owen Johnson wrote in a book for boys, "The theory of taxation is to soak the taxed all they'll stand for, but to leave them just enough so they'll come back again." The theory has been in use as long as governments have been around. If the new stumpage rates hadn't caused screams, they would have been—by definition—too light. If the new rates had caused people to get out of the industry, they would have been too heavy. The proof of the reasonableness of the new rates is that they've been raised since then, not lowered.

Williams was right in stopping chip profits from being untaxed gifts, but he was wrong on two counts. His timing was bad. Raising taxes when profits are going up minimizes the amount of screaming and lets people get used to them before the market goes down and they start to hurt.

And he didn't really change the stumpage system; he modified it—just as twisting the nozzle on a garden hose changes the pattern of the water. He continued to incorporate a manufacturing tax and a raw materials charge into a single payment to government. As long as the two were lumped together, industry could continue to complain about high taxes. If there had been a raw materials charge separated from an added value tax, the public, and Williams's fellow politicians, might have better understood where the money came from.

Harry Gairns and Howard Lloyd—the future Social Credit MLA for Prince George—understood this. But they argued that since forests were a renewable resource and the industry was being charged for reforestation, the province didn't have the right to charge for the use of its forests.

Of course, unless the government receives an incentive—revenue—it has no reason to allow trees to be cut. And as both Federal Government and private studies point out, wood costs less in British Columbia than in other provinces and far below what is charged in many other countries.

While this argument bored the public, lumber prices kept dropping.

On January 12, 1974, Bob Williams appointed a three-man Task Force to look at the old "temporary" tenures—the ones on the Coast that said a company could keep paying a low fixed rate as long as trees that were worth cutting remained in the area. Peter Pearse, ex-Liberal Federal candidate and consulting economist in his time off from teaching at the University of British Columbia, headed the group. The other members were the province's Chief Forester, Ted Young, and a respected industrial forester, Bill Backman.

On February 6, the government purchased Kootenay Forest Products.

At the end of February the Task Force recommended sweeping changes. The industry screamed that the task force recommendations violated "the sanctity of contracts"—as the industry had often complained in the past. The industry implied that unless the government kept all old agreements in force, with all implied provisions and accepted practices, all business would flee British Columbia feeling that the government was untrustworthy.

The Task Force had anticipated this reaction and argued that just because the government had been foolish in the past, it didn't have to continue to be. Being robbed yesterday was no reason to accept being robbed today.

Lumber prices kept dropping.

Labor problems kept increasing.

In September 1973 the BCR workers had gone out, then returned to work so they could go out again the next month. When North Central Plywood opened its doors, it had quickly closed them because of a strike over safety. There had been a total of four rail strikes in 1973, and it looked as though 1974 was going to be even more of a boom year for selling picket signs. On July 10 workers at Intercontinental Pulp and Prince George Pulp walked out, joining Pulp and Paper Workers of Canada members at seventeen other mills in a wildcat strike. The industry was supposed to be negotiating a new contract with the IWA, but nothing was happening. John Ernst, who owned the Quesnel newspaper, published a full page ad listing the IWA's demands—something both sides had agreed not to publicize. The ad ended with this statement: "Due to the above, Ernst Forest Products will cease purchase of logs after July 12th." On July 14, Bill King, Minister of Labor, got the pulp workers back on the job, for the time being.

On the twenty-fourth, the Bank of Canada raised its prime rate to a record nine and a quarter percent. Housing starts in British Columbia dropped to a postwar low. Mortgage rates went up in the United States. Lumber prices had dropped $61 per thousand board feet of two-by-fours since the previous February. Mills began laying people off.

On August 8, President Nixon became ex-president Nixon, and the next day the Task Force released its second report. It said that the Vancouver Log Market was rigged, and the government was basing its Coast royalties on manipulated figures.

By the end of the month, 1,777 woodworkers in the Northern Interior and Cariboo had been laid off. At the beginning of September the pulp workers went on strike, legally. On September 29, Clear Lake Sawmills closed down "indefinitely." The market for lumber was terrible, and the sawmills couldn't sell chips to pulp mills closed by strikes.

By October 8, 2,429 woodworkers in the north had been laid off. In the States, U.S. Plywood closed its Seattle mill. Twelve thousand sawmill workers in Quebec were out of work. On the twenty-fourth, Ernst Forest Products closed down, Weldwood laid off 125 employees, and Triangle Pacific gave another 90 the bad word.

That day the owners of thirty northern sawmills formed a group called the Independent Chip Producers. They claimed they were being cheated on the price of chips, which made up a third of the volume of timber they processed. "If the pulp mills would pay a fair price for our chips, then all those men would be back at work right now," their unnamed spokesman told the Prince George *Citizen*.

Marginal firms went bankrupt, followed by businesses that sold things people could do without, like sporting goods, records, and gifts. Prime retail space was available in Prince George for only a third of its alleged rental rate.

On the twenty-eighth, Ian Mahood, president of the Truck Loggers' Association, met with Premier Barrett and Bob Williams, seeking their help. The same day a new organization, the B.C. Independent Loggers' Association, also met with Williams. This new group, the amalgamation of regional groups representing people who owned and operated their own logging trucks, skidders, and other equipment, had trouble with contractors who failed to pay, as well as trouble finding work.

Bob Williams had been handed an opportunity. Millowners were saying in the press that their utilization standards were so low that a third of each tree went for chips. They were calling on the government to take action, instead of resenting any hint that the government might interfere with free enterprise.

He could have allowed companies to go broke. After all, management at many mills was inept, the government did not owe millowners a living, and the "proof" of free enterprise is that bad companies are driven out of business by better ones.

He could have legislated a chip price higher than the pulp mills had been paying.

He could have said that the pulp mills were conspiring to keep chip prices down and taken them to court.

On November 14, the Timber Products Stabilization Act was introduced in the legislature. Williams told reporters that the purpose of the bill was to set minimum chip prices.

Williams later recalled:

> *Pearse (in his Task Force reports) wanted a timber authority. And when the industry went into decline in 1974, we brought in the Timber Products Stabilization Act. We gave it that title so it wouldn't scare people, even though it was really a timber authority. People were never able to grasp it, to use it for headlines. It was a major intervention that gave sawmillers a fair price for chips. Great numbers of sawmillers survived because of it. It was a move to redistribute wealth from pulp mills to intermediate sawmills.*
>
> *In establishing a chip price we had Pearse and the B.C. Research Council work on it and talked with industry. We didn't rush into it.*
>
> *Chip stabilization was only the first step. The Crown didn't receive any royalty on residuals going into pulp mills. Those residuals are the same as logs to a pulp mill, so why not get some rent on them? The idea of the act was to use that rent for sawmills in bad times and in other market periods the idea was to give some cash flow to the Crown. Eventually we wanted to establish a Forest Products Board to intervene in the log market on the Coast, which—as Pearse explains—isn't a free market.*

On November 19, layoffs totalled 15,258 according to the Council of Forest Industries.

On the twentieth, the newspapers seemed to prove that Williams's estimate of the industry was at least partially right. Gary Bowell, president of Weldwood of Canada, used the standard line about destroying the industry's confidence in the sanctity of contracts. The Opposition in the legislature worried about the vagueness of certain clauses. The independent sawmill owners didn't say a word.

The pulp mills raised the chip price before the bill was passed to try to get the public to think that legislating minimum prices wasn't necessary. When the bill became law on January 15, 1975, the pulp mills said it was, of course, redundant.

The lumber price stopped slumping. It was flat on the floor.

Northwood announced it was going to close down Eagle Lake Sawmill at Giscome. The BCR went on strike again on November 21.

The story of the year was summed up by Rim Forest Products in Hazelton. The mill had burned down in February 1972 but had been rebuilt for $1.7 million. Within five weeks after the fire the mill was producing 50,000 board feet per shift. "We tore it all out, salvaged what we could and rebuilt it," Jack Ellett told *B.C. Lumberman*. "We needed something fast and we built something fast. We didn't bother with blueprints, just drew pictures on the side of the burner." At full production the new mill turned out about 115,000 board feet a shift.

Then came the slump. In July 1974 the firm closed down. For the next five months the owners blamed both poor markets and the NDP government. On August 2 the firm went into receivership. Through the press, Rim's owners kept a running battle going with the B.C. government. Williams avoided them at times, having learned from Dave Barrett's experience with northern egg producers that "confidential" meetings often end up reported in the press. He suggested alternative methods of getting the mill going again—none of which included the owners who had tried to blame him for their failure.

At last, in November, serious talks were held. On December 1 Rim re-opened, thanks to refinancing by the Royal Bank and a reduction of stumpage rates to the minimum by the government. The main product was chips for the government-owned CanCel mill. "Lumber will be a sideline product for a while," Bill Sterling said. "We are getting a decent price for chips, and with the minimum stumpage rate in effect, we should be able to operate profitably."

The new year, the last year the NDP was in power, opened with an article in *B.C. Logging News*, by Gray Wheeler, who attempted to step back from the moment, from the times he was part of. He gave a remarkably objective view of Bob Williams (the bold italics are Mr. Wheeler's):

> *...His public posturing in the past, which may well have been the defensive posturing of a basically shy man, has added to the aura of ruthlessness surrounding his administration. Whether he is actually ruthless or not, the important thing to British Columbia is the fact that the Forest Service and the forest industry believe he is ruthless and vindictive. Neither is functioning effectively and both are contributing to a profoundly unhealthy situation....*
>
> *Of the significant changes he has wrought during his two year term, none in themselves has been enough to cause a mass exodus of forest industry capital from the province. In the long run our economy is sound, given proper resource management....*
>
> ***With the exception of government takeovers and the hurried implementation of the Logging Guidelines, Bob Williams has not bowed to political ideology or lived up to his own public rhetoric. For a purported socialist he has proceeded with a degree of conservativeness in most matters and deference to research....***
>
> *It is time for Bob Williams to pull the curtains on his giant-killing act and settle down to some serious and sincere dialogue with his industry. It is also time for the management of the forest industry to pull their heads out of the sand and forget the past. It's a new ball game—a game which is changing all over the world....*

For the forest industry and the government, it was the bottom of the ninth inning in a scoreless game with two men out and no men on and the pitcher coming up to bat. Contractors, like Twin River, went under. Northwood stopped logging operations a month earlier than usual. Lumber stockpiles buried some mills. The residents of Giscome were given notice that the mill would be torn down and that all those who rented would have to leave by the end of June. A year and half later Giscome was a straggle of fading houses along the road. Mallards nested in the slough that ran through the middle of town.

On March 18 the Nazko and Kluskus Indian bands closed their lands to the government and to industry. Their blockade closed off two watersheds with timber that mills in Vanderhoof and Quesnel wanted. Three days later Lakeland's sawmill burned down, just a month after it replaced a planer that had burned down the previous year. Mills unable to sell lumber chipped good wood, but the pulp mills weren't buying. Instead the pulp mills were surrounded by mountains of chips that turned yellow-brown with age. At the beginning of May Northwood announced it was laying off 1,200 more workers.

Good news, like Eurocan's announcement of two new sawmills in Houston, the success of Babine Forest Products, and the naming of Peter Pearse to head a Royal Commission on forestry, couldn't compete with the bad news.

Pulp prices had dived when a new Swedish government dumped five million tons of pulp on the world market. The previous Swedish, socialist government had been buying pulp for three years to support prices, and the new government was out to "prove" this practice was wrong. Since the northern B.C. pulp mills had been designed to produce Scandinavian-style pulp, at a lower price, they couldn't sell a thing.

At this point, pulp companies knew a strike in the industry would save a fortune. They wouldn't have to pay workers, and the poor market meant the firms could afford to sit idle, draining strike funds so the unions wouldn't be able to afford a strike when the market improved. The poor market would let the companies force a settlement that offered workers a loss instead of a gain. It would be wrong to say that the pulp companies bargained in bad faith. When it came to negotiating, they were atheists. On July 13 the Canadian Paperworkers Union and the Pulp and Paper Workers of Canada went on strike. On the twenty-third Northwood laid off 700 woodworkers, and management in the industry threatened to close other sawmills.

In Squamish, the railcar plant the government had started was at last producing lumber cars, but there wasn't a market for lumber and there was no need for chip cars during the strike.

On August 15 the Nazko and Kluskus Indian bands ended their blockade, dropping their demand for $7 million, in return for the government's promise of "meaningful negotiations." A week later the BCR was closed by another strike, so lumber couldn't be shipped from the north. Again. On September 9 a labor dispute closed all Vancouver area supermarkets. On the same day, IWA members who wanted to get to work crashed the line of "information pickets" from the Pulp and Paper Workers of Canada in Fort St. James. Two days later the same thing happened at Bulkley Valley Forest Industries.

On October 6 the government brought in and passed the Collective Bargaining Continuation Act, which sent workers in the pulp, food, and propane industries back to work as well as returning BCR workers to their jobs, for a minimum of ninety days. Some union leaders said the act was betrayal. The NDP said that many of the same unionists who complained had told the government, "Look, we're caught in a bind. If we end the strike now we'll gain nothing and our members will feel we've misled them, but there's no way we can negotiate a good settlement. Get us off the hook." Whatever the truth of the matter, a lot of hungry people were glad the strike had ended. On the ninth the IWA signed a tentative contract agreement. Then the teachers demanded a high settlement and the post office workers demanded a seventy-one percent pay hike.

In November Dave Barrett called an election. On December 11 a new government was elected.

Three years passed.

It was raining in Vancouver, of course. Bob Williams was willing to talk about his experience as Minister of Forests. After hearing about him from members of the forest industry, one expected a sinister figure in a sterile office punching computer buttons and making small animals scream. Instead, Bob Williams's office was in his home. The kitchen looked as though someone in the family enjoyed cooking. Toys were scattered in the hall. His office was a jumble of books on chairs, on the desk, covering the walls. When asked a question, Williams often paused to look out at the gray day and the seagulls, the log rafts and the water, before replying. Throughout the interview his voice was gentle.

Question: *There was a great deal of concern that you were going to have the Forest Service take over logging. It was NDP policy, and you had the Service do some logging for Plateau Mills.*

Bob Williams: *The idea was to educate the Forest Service. When the Forest Service first developed environmental guidelines, the industry was right: they were costly, wasteful. But we had no measuring stick to compare with. I wanted to bring up the level of understanding in the Forest Service and avoid future mistakes like that one.*

Question: *What about the old game of companies saying, "We just built a bigger mill, we need more timber or we'll have to lay people off"?*

Bob Williams: *We said, "You won't get an increased cut no matter what you do. You're free to build a new mill—if you want an empty sawmill." The Forest Service has wanted licenced mills since the year one because they don't understand competition. I feel the real fight in B.C. in economics is the sharing of the pie between workers, the companies, and the government. Wherever you look—even at the way wages can be offset against stumpage—you see a common interest in screwing the Crown. And thus far the Crown has always ended up short. We have to open up the system, to bring competition for timber back. It's the most necessary thing in the province today.*

Companies are grossly incompetent and everyone—at every part of the political spectrum, even conservatives—should worry about that. Pearse recommended gradual phasing in of competition. At the time I thought it was a good idea. But today I'd do it cold turkey, like the change from Fahrenheit to Celsius. We are running out of genuine entrepreneurial ability in the province. In the past it was often the small guys who made changes, but I'm going to repeat this: **we are running out of genuine entrepreneurial ability in the province.**

Question: *What can be done to change attitudes on the industry side?*

Bob Williams: *It's not that at all. It's a matter of making them sink or swim. In some parts of the province it would be easier to do than in others, so you'd have to do it by area.*

Question: *What about competition under the present system? Isn't there room for people to get into salvaging timber, for example?*

Bob Williams: *Companies in recent years haven't invited people in. In fact they've gone to great lengths to keep people out of the areas they're in.*

Question: *There was talk of a timber marketing board. Lumber output is too high at times.*

Bob Williams: *Lumber is the closest thing to a genuine classic competitive output market in the North American economy. There's no need for intervention in that sphere. However, we did intervene to maintain production in times of a poor market. I think the state is justified in keeping employment up.*

As to whether the competitive output market is allowed by companies because they can bind their risk on the production side, I feel we could open up competition without harming the output market.

Question: *What about the political realities, the real world of caucus and voters affecting actions?*

Bob Williams: *I spent a lot of time covering the tracks of others, but I was relatively free. The blame for things undone has to rest with me.*

If I had to do it again I'd buy four years full bore— making needed changes. But you can only do that if you know where you want to go, and I don't think many MLAs know what they would do with the job.

Question: *There's a variance in the NDP between doing things that are politically acceptable and doing things "because they should be done." In terms of company money that could be poured into fighting change, would you do what you felt was right if you were back in, or try for something like a trade-off? Getting rid of property taxes in return for reforming the forest or suchlike?*

Bob Williams: *You just have to be naive enough to do it. You just have to go ahead not knowing or not caring about anything other than the simple fact that we have to give the forest back to the public it belongs to.*

Question: *Could you do it?*

Bob, smiling: *I'm naive enough.*

Question: *Could this be sold to the public?*

Bob Williams: *I doubt it. The public isn't interested.*

The talk turns to other things: the Environment and Land Use Committee, Not Satisfactorily Restocked Forests, the Pearse Commission. It's getting dark. It's time to go.

Question: *Throughout all this you've talked about things undone, and I'm the one who's brought up your accomplishments.*

Bob Williams: *There's too much ego in politics. It's one of the things I'm glad to be away from. However, I'll say this: I did all a human being could do in the time allotted.*

Back in 1513 a fired civil servant told the whole story when he wrote:

It must be considered that there is nothing more difficult to carry out, nor more doubtful of success, nor more dangerous to handle, than to initiate a new order of things. For the reformer has enemies in all those who profit by the old order, and only lukewarm defenders in all those who would profit by the new order...partly from the incredulity of mankind, who do not truely believe in anything new until they have had actual experience of it....

Bob Williams, the unions, and the forest companies simply proved Machiavelli was right.

Chapter 24
Catechism on a Jigsaw Puzzle: The Fourth Royal Commission

Why are Royal Commissions needed? Are they rigged? Are they biased? What are they really like? Do they have any effect? The 1975-76 Royal Commission on forestry is recent enough to be taken apart piece by piece to see what happened, and what could have happened but didn't.

Why was a Royal Commission needed?
According to Bob Williams:

> *If I'd gotten rid of quota—taken rights away from companies—the shit would have hit the fan. The IWA would have joined management in saying jobs were in danger. I had to keep the machine running while changing its direction. It was a darn hard problem to wrestle to the ground.*
>
> *I wanted a consistent, logical package so the industry wouldn't be scared out of its jockstraps. I was satisfied that I alone couldn't unravel the industry with its complex tenure problems and established practices that weren't even mentioned in legislation. So I set up two task forces to solve immediate problems and to evaluate Peter Pearse. His task force work was very good....*

According to Peter Pearse, the man Williams chose as Royal Commissioner:

> *I was concerned that allocation of timber was getting out of hand. There were a lot of **ad hoc** arrangements not firmly grounded in law, and the whole situation had been evolving very quickly. The pace of consolidation was very fast, and I believed it was taking place unwittingly, especially Tree Farm Licence Allowable Annual Cut expansions without any new allocations. I was concerned about declining competitiveness and diversity, and I was concerned about the way they figured out the allowable annual cut.*

In addition, it had been almost twenty years since the last Royal Commission, and times had changed. The last goal the Forest Service had been given to strive toward was sustained yield. The rules were suitable for an industry that had disappeared. Williston's insistence on flexibility meant that little that the government, or the Forest Service, did was in law or regulation.

Instead the government's strategy and aims existed only in the mind of the Minister of Forests.

Was the Pearse Commission rigged?

Bob Williams said, "I discussed my concerns with him—how I felt—before he began work." According to Tom Waterland, "No. I had lunch with Dr. Pearse once or twice so we could get to know each other, but I made it a point not to discuss it." And, he said, "There were five items in his terms of reference, and I agreed with four."

When Pearse was asked if either of the ministers ever said, "Hey, here's what I want to see in your report," he replied:

> *Unequivocally no. The new minister had a completely open mind. And, as you know, Bob Williams criticized some parts of the report vigorously. Neither government tried to exercise its influence on what I wrote, nor saw a first draft. I am a very independent person and wouldn't take kindly to someone telling me what to say. Bob Williams obviously had confidence in me, or else he wouldn't have had me appointed.*

This *sounds* like the usual political poppycock. Royal Commission reports have been rigged. Both the NDP and the Social Credit reports on redistribution of electoral district boundaries were rigged, according to people who quoted conversations from meetings of MLAs and executive assistants busy gerrymandering district sizes and shapes.

However, the Pearse Commission wasn't rigged. Williams set up the ground rules: the things the commissioner was to investigate. Williams was asked, "Did you input Pearse or guide him?"

> *After his testing on the task force I felt he viewed public interest in a way compatible to myself, so he was freed of his two colleagues.*

Question: *That didn't answer the question.*

Bob Williams: *That's right.*

Question: *I know it's not nice to bring these things up, but what did you do with Pearse?*

Bob Williams: *I discussed my concerns with him—how I felt—**before** he began work. I certainly hope you'll ask Waterland the same question.*

Pearse pointed out another reason Williams didn't keep watch like a mother hen. The terms of reference, he said:

> *...were unusual in an important respect. They specified not only the issues to examine, but also the objectives toward which I was to draft my recommendations. I didn't want to have to make political judgments—the share of the value of the timber that should be received by the public rather than accrue as industrial profits. I wanted the government to say whether they wanted to collect full revenues or not. They said, "Yes, we do." I didn't want to decide independently whether companies should be charged the full value of their raw materials or not, because that's a value judgment....*

In other words Williams felt secure because Pearse was equally concerned about the concept of "full economic rents." Waterland looked over the terms of the reference and probably felt he could pick and choose things that would fit his needs, much as Williston had done with the previous Royal Commission report. And Pearse made it clear he wasn't going to be anyone's henchman. The report reflects this determination. And a lot more about Peter Pearse. Which brings up the next question:

What about bias?

Objectivity is a goal rather than a result in a work of any length. The Pearse report reflects the fact that Peter Pearse is a professor. But he's more than just a teacher with a doctorate. Pearse has been a partner in an economics consulting firm, a candidate for the Liberal party, and a timber cruiser for B.C. Forest Products.

Seated in the Faculty Club at the University of British Columbia, Peter Pearse presents a picture of an Oxford Don cast by Hollywood: Harris tweed jacket, salt and pepper hair, a precision of speech uncommon in any walk of life. He carries himself with the assurance of a captain of industry but smiles a lot, as though to reassure the person he's talking with that he isn't out to hurt them. And he speaks freely about his own biases.

> *I had some ideas about taxes and charges, but not about export policy. I had no preconceived ideas about competitiveness of intermediate resources—chips.*
>
> *The idea that the government should collect the full value of its resources is a new idea. That it was in my terms of reference was a great break with tradition. Fish, game, range rights, water, minerals have all historically been given away in this and other provinces, or made subject only to very modest public charges. Governments are now shifting to charging for their resources. Alberta, of course, now has very heavy royalties on Crown oil and gas. And now even the Federal Government is talking about charging fishermen royalties on the fish they catch. But the government couldn't legitimately take away what had already been alienated....*

Question: *On page sixty-six of your report, talking about competition, you sound a bit like Adam Smith.*

After an hour and a half of conversation, Pearse's voice shows emotion for the first time: a bit of righteous anger, an outpouring of belief:

> *Perhaps I'm basically a nineteenth century individualist. But I believe that, in general, competitive markets are more in the public interest than monopolistic ones. That's why enlightened capitalistic countries have competitions legislation and anticombines laws, and there's a real need for government to maintain a vigorous competition. The stronger the monopolistic tendency, the stronger the case for government intervention. Look in my report. What you can find is recommendations to encourage competition for chips, logs, and some timber. But, being realistic, not all timber—just a little. I suggested a portion being reallocated, but didn't really suggest wide open competition except for Timber Sales Licences. A very modest proposal.*

Were the proposals given to the Commission equally "modest"?

The B.C. Independent Loggers' Association (BCILA) wanted twenty percent of the cut in the province given to independent loggers on the basis of "performance, financial capacity, and demonstrated stability in the industry." In other words, a bigger share of the action with no chance for newcomers to get in on it. They also asked that "game corridors, fire breaks, burned areas, all types of reserves including parks, and all existing licences be kept free of blow-downs and dead and dying trees," with the right to harvest going to independent loggers. In the north, ten to fifteen percent of a normal forest is blow-down. The majority of the timber in the province is past the peak of its growth cycle, "dead or dying" in their terms. Following the BCILA modest proposal would have wiped out every forest in the province. The association's brief continued with a few more requests: that salvage logging after fires or windthrows, or as a result of insects or disease, should be allocated by local rangers instead of going to competitive sale. The BCILA then suggested that the possible expansion of parks or other reserves that might slow down harvesting should be guarded against.

Another group recommended to Pearse that "an immediate study (by an independent body)* be carried out to determine the present status of our forest resources...." Perhaps a Royal Commission on Forestry?

Was Pearse able to maintain a straight face during all this?

Usually.

That was because most of the briefs submitted and people who talked at hearings followed the same pattern:

1. These are the only facts that really count.
2. They show that our view of the forest industry is the only correct one.
3. They also show we are getting a raw deal.
4. Therefore you should do everything possible to help us.

However, at times Pearse enjoyed himself. Especially when he could trip up a sharp-witted opponent. It was the duel of wits he enjoyed, not "proving" someone right or wrong. When Canfor undertook to defend its brief, the transcript of the hearing shows this:

Bentley, Patterson, Macaulay, Armstrong Smith
Questioned by Commissioner
Q. On page 6 of your brief, you say that, "Typical of sawmill development in other Interior locations, it..." being the Chetwynd division, "was the ingenuity of the smaller independent operators combined with the strength and financing capabilities of the larger integrated companies which permitted centralized operations to become established." How typical is it that the larger integrated companies' help has been a feature of sawmill development in the province? My impression was that it was not a typical feature.

A. I think it's a play on words and the word "typical." I think it has been a typical development that smaller independents emerged as what would have been giants at the time but which now are accepted as reasonable sized mills, and I quite follow what you are saying Mr. Commissioner. There are some cases where integrated companies have acquired former small bush mills but there are more instances where smaller bush mills have been consolidated and a larger entity has emerged as a centralized mill with full recovery potential including chips.

---

* Parenthetical remark is part of original quote.

> Q. The connotation that I put on this sentence is that an essential element in the Chetwynd operation is the ingenuity of smaller independent operators. Now what role do they have in that division now?
>
> A. There is no on-going in operators in the sense of mill people, Mr. Commissioner. We do work essentially, including phase contractors, a hundred percent with contractors in Chetwynd area....

In response to both questions, Canfor had contradicted its brief. The situation got worse a few minutes later:

> Q. ...your conclusions are based partly on your experience and partly on your firm's surveys, your surveys, do you consider them to be more thorough than those of the Forest Service?
>
> A. Well, I don't like to say that, but—
>
> Q. Go ahead if you feel—
>
> MR. BENTLEY:
> A. Glen is a professional forester, Mr. Commissioner, I am not. I am convinced that ours are more thorough at this stage of the game. There is no suggestion here that the Forest Service don't have every talent and opportunity to do the work, it's a question of priorities.
>
> MR. PATTERSON:
> A. We could certainly, Mr. Commissioner—
>
> Q. Not being a forester makes you less inhibited but does it make you better qualified to make that judgment?
>
> (Laughter)

Three years later Pearse broke out in a grin when the Canfor discussion was brought up and said that it had been a most enjoyable session.

Why did people send in briefs or appear at the hearings if the possibility existed that they'd be taken apart?

On the whole, when people weren't asking for the chance to buy Crown land and be forever exempt from taxes—as Canfor did—when foresters weren't arguing with each other, when people weren't worshipping sacred cows or roasting them, the hearings were gentle. And the participants were able to change Pearse's view of the industry. They often presented valid points about "raw deals" even though briefs often directly opposed each other. They demonstrated the complexity of the industry and of the issues facing the commissions. When all the briefs were stacked together, from forest companies, government agencies, Indian chiefs, municipal councils, and environmental groups, as well as individuals, it was made obvious that simplistic solutions wouldn't work, that there was no single true path to justice. By presenting differing views, they helped assure that the Commissioner wouldn't be swayed by a single pressure group.

So how did Pearse handle all this?

Exhaustively. However, he was wise enough to ignore some of the things he was asked to do, give short answers on others, and suggest a separate Commission on the Forest Service. He used his research staff to check out the past so he could write a very good history of how tenure systems and the industry structure evolved. Then he concentrated on the things he felt were essential—for 381 pages of small type, plus a second volume with appendices.

The reason he didn't cover everything he was asked to cover is simple. No one could. He was asked to look at how forest harvest rights were given by the government, what the effects of past policies had been, whether the money coming to the government from the forest industry in the past had made sense, how the money was collected at the time, and to detail the structure of the industry. Then he was told to look after the public interest by making sure

> *the full contribution of the forest resources to the economic and social welfare of British Columbia is realized...(and that)...the various public levies on, and charges associated with the acquisition of Crown Timber reflect the full value of the resource....*

In his spare time he was to make recommendations to assure "the efficiency and vigour of the forest industry" and suggestions to assure the forest would be properly managed.

And the result was?

Long. Pearse had a lot of ground to cover, so much that he set himself five goals and did only a token job on everything else. The goals were:

> *...clarification of resource management goals; articulation of a deliberate policy for the pattern of industrial development; improvement of the security of timber supplies provided through the tenure system; enhancement of the scope for government flexibility in the allocation of rights to public timber; and the development of improved structures and procedures for the administration of forest policy.*

He discussed these things, but a lot of the talk was an attempt to get at two other problems—one mentioned in an almost unrecognizable form in his goals, the other not mentioned at all. He wanted to find a way to give the Crown "full economic rents," and he wanted to put competition back into the industry.

The government's "charge"—the statement of what he was supposed to do—said, in effect, that he should make sure the government wasn't giving away its timber so cheaply that companies would make bigger profits than were "fair." Who's to define what's fair? In his checking out of companies, Pearse found that *according to the annual reports available* the firms were making a poor profit. These reports excluded companies formed to supply other firms with low cost goods, like pulp companies formed to supply their owners who made paper in other countries. The reports excluded family-owned firms. They excluded a lot of financial juggling. As one industry executive said, "Sure we make lousy profits some years. But when we make up for it, we make out like bandits! Sure there are a lot of firms with low profits. But look at the ones big enough to be efficient, small enough to avoid executive ineptitude." And, of course, the major profit a firm in the industry can make is selling out, changing its tenures into cash.

In addition, many of the companies appearing before Pearse pointed out that not all benefits were cash. A new plant that means more jobs—or an old one that keeps people employed—is of benefit to the province. Building a main haul road that will be used for decades costs a lot and is offset against stumpage payments, but the Crown benefits because without it timber couldn't be harvested and provide revenue. If firms were to go the cheapest possible route, they'd go through the forest like mechanized locusts, cutting a continuous path. Cutting in small blocks drives costs up, but it's better for wildlife. The companies argued that these are also economic rent to the government.

Pearse had the opportunity to recommend a definition of "full economic rents," a phrase that seems to change meaning with every pair of lips that utters it. He had the opportunity to suggest a yardstick to measure what was happening and what should happen to the forestlands of the province. He didn't. Instead he concentrated on rates of harvest, changes in regulation, changes in taxation. Martin Buber, the theological philosopher, divided religion into three phases: creed, a basic statement of belief; code, the laws of conduct; and cult, all the trappings that gradually envelop any religion. In the forest industry, in a British Columbia that has continually worshipped the forest as the source of its jobs and its revenue, Pearse spent a lot of time on cult, touched on code only in one place, and avoided creed.

Where he rose above methods of calculating the allowable annual cut and security of tenure was in his approach to competition. Here he sought to change the law:

> *...many mills of large corporations, both sawmills and pulp mills, are well beyond the size that most experts consider necessary to achieve production efficiencies.... In saw-milling, most observers agree that some of the most advanced, innovative, and efficient mills have been built by small companies. Moreover, I have found no evidence to suggest that, even among the large integrated operations themselves, either technical or economic efficiency is correlated with corporate size....*
>
> *In my opinion, the continuing consolidation of the industry and especially the rights to Crown timber, into a handful of large corporations is a matter of urgent public concern....*
>
> *The efficiency of private enterprise depends on competition not only in product markets, but also in the markets for the input of capital, labour, and raw materials. The argument that competitive sales are disruptive is only partly true, and disruption is often confused with sometimes painful adjustments in an industry that are necessary to maintain its vigour and efficiency.*

Scattered from pages sixty-two to sixty-six of the report are these beliefs, ideas that Pearse returned to again and again. In the sections quoted, and throughout the work, he gave a thorough diagnosis of the problem—from the viewpoint of eighteenth century individualism. Later he was asked, "What about competition as an idea? Plastics companies don't compete for raw materials. Neither do a lot of other....And there's no way the mythical man with six strong sons and little capital can get into the business today." Pearse replied:

> *There is certainly the possibility of making arrangements to enable new entrants to the industry, and there's an important role for the small operator. We still have a few small licencees. Admittedly the cost of entering the lumber industry has increased and the hand logger is obsolete. But there are a lot of large and small contractors. However, that isn't the important point. The government doesn't have to put extra obstacles to entry to the industry through their tenure arrrangements, as they have.*

Today you need $10 million to $20 million, plus timber rights, to get what most industry people would call an efficient mill going. The industry estimate is based on the common practice and is therefore high, since smaller, specialty mills can show handsome profits. But a smaller mill ends up paying more for materials and more for the money it borrows. Changing tenure, "opening up the forest," is indeed part of bringing competition to the industry—but only part.

If a man had the money to build an efficient sawmill, he'd have enough money to live well without working. The timber rights needed to keep the mill operating cost so much that if a man could afford them, his children and their children could live well without working if the money were put into Canada Savings Bonds instead of into a sawmill.

To open up the industry a bit and to finally get into law what had been going on for years, Pearse suggested that Timber Licences should be modified and given a new name, Forest Licences. These would have ten-year terms, after the first term, which would be fifteen years. To allow planning and road construction, a firm could ask for renewal bidding three years before the expiration date. Instead of suggesting that firms be guaranteed automatic renewal, as the lumber and pulp companies wanted, or fully competitive bidding at the end of each licence period, as some people wanted, Pearse took a middle course designed to put a measure of competition back into the industry. He suggested that firms be guaranteed sealed bids for eighty percent of their present timber, with additional timber going into the pot if the firms spent their own money to improve the productivity of the forest. Productivity could be improved through, among other things, planting, fertilizing, spacing young trees by thinning after a few years, or trimming the trees as they grew.

This suggestion raised the blood pressure of the industry. Soon a standard argument, heard from Bob Wood, Lakeland, and the Council of Forest Industries, among others, began to be advanced against Pearse's scheme. "Look," they said,

> *let's assume there's a medium-size block up for auction. Now that might be 100 percent of a firm's needs. But it also might be just 10 percent of a big firm's needs. The small firm really wants that wood. But the big firm can spread any extra cost in getting that block against the 90 percent they got for a low cost as part of normal renewal. Who do you think is going to win? The guy with the need, or the guy with the bucks and the ability to spread the cost of reaching an efficient level of production?*

Their argument is correct, as far as it goes. But in the United States, for instance, the big firms can't compete in some sales. Pearse suggested— vaguely—that there could be safeguards. One form he considered, but didn't spell out, would be to create three, or five, or whatever number was necessary, different classes of sales ranging from those that were wide open to those open only to very small, small, medium, large, or giant-size operations, with a selected size competing for each Timber Licence. If only small or medium firms could compete at a given auction, the medium firm, or a small firm with a bankroll, could win. This provision would allow changes in operation size, through competition.

But Pearse didn't just want competition for the tag ends of present Timber Licences. He suggested changing from an area licence to a cubic-foot or cubic-meter licence. So instead of getting an area and either finding that it didn't have all the cubic meters of raw material bid on because of a poor survey or discovering more wood than was allegedly in the sale, each firm would be guaranteed a certain amount of wood. This guarantee would eliminate the problem of cruises that were often off twenty percent or more. Every time the cruise erred by underestimating timber, the Crown would have more to auction off. Further, company fiefdoms would be eliminated, since the Crown would say what should be cut and where in an area bigger than a PSYU. So management would be based on the needs of the forest rather than a company's holdings. This provision would also let firms get their wood from as close as possible to the mill, instead of passing by three mills on their way to a patch of forest that was their "interest area."

And Pearse suggested that the idea of forcing bidders to put up forfeitable deposits was obsolete.

Although Pearse wanted to encourage competition, it is almost impossible to force people to compete. The industry phrase "disruption of planning stability" is used by monopolies, cartels, and oligopolies. To the forest industry, competition is an unnatural act.

If the government were to stop faking free enterprise and tell the public how much it controlled the forest industry, the Loyal Order of True Believers in Free Enterprise would be outraged. Immediately afterwards, it would dawn on the public that if the government does indeed control the industry, then everything the public thinks the industry does that's bad is Victoria's fault. An outcry of delegations would pack the corridors of the Parliament buildings. The industry would scream in anguish as new regulations were imposed.

If the government used its power to increase the chances for genuine competition, the results could be just as drastic. Imagine the screams if firms were told, "You can be in lumber *or* pulp, but not both." Or if a Forest Service group went around taking timber back from companies that violated their agreements with government. Although firms have overcut, one cutting five years' allowable cut in two years in a recent case; although there are at least two Tree Farm Licences that could be reclaimed; and although Forest Service officials say that a number of timber licences could be revoked under the law—little is done. A few "third band" licences have been revoked. Some timber that companies decided was junk has been taken back. That's all.

Both of these approaches are too radical to be politically acceptable, and Pearse realized this. Instead, he suggested changing the ways timber is disposed of by the provincial government to allow the possibility of competition.

Unfortunately, although it's obvious in conversation with him that Pearse realized the limits of governmental action, in his report he ignored the need for public pressure if the government was to make changes.

The people of British Columbia don't think about the forest industry. Every time a politician says, "The public demands...," he lies. Families getting together for a social evening in Burnaby, Hope, or Mortgage Heights in any part of the province do not discuss governmental policy and company structure, competition, and enforcement unless the host thinks the hour is late and wants people to leave. The industry exists. It goes through good times and bad. It provides jobs. It often complains about government changes that the public doesn't understand or care about. So what? Pearse had a small opportunity to point out to the public what was going on. He didn't. Later he said:

> *The realities I had to consider were what would best serve the province AND how to bring about the needed changes. I was very concerned about practicalities of industry and government commitments, but not political*

> *realities. I worked very hard to make sure each of my recommendations was practicable.... When I finished the report I told Tom Waterland it would not be proper for me to engage in public debate until the government made up its mind about what to do....*

That was a mistake. Pearse did an excellent job of analyzing the problems in the government and industry, but he didn't provide the pressure needed to make sure his recommendations would be adopted. One of the few forestry reporters in the province said of the report:

> *Read it? Are you kidding? I've looked at it, read chapters, but there's just too much. How many recommendations did he have—five hundred? It's been three and a half years since the report came out and I'm still discovering new things in it. Trying to review that thing the day it came out was like trying to do a quick review on a new book called the Bible.*

When the report was released the press called it long, thorough, and "interesting," but the public's main source of information about it came from critics because the press didn't wade through all Pearse wrote, and the public didn't bother.

Later Pearse seemed aware of the problem caused by doing such a sweeping job. "One question reporters often ask is whether, if I had a charge to rewrite the report, I'd change it, " he said. "I don't know whether I'd change much, but I'd argue some things differently."

Unlike Sloan, Pearse *cared* about the forest industry and what the people of British Columbia got out of it, so he did an excellent job. For example, he showed that the U.S. Forest Service, which administers about the same amount of timber and land, had ten times the staff the Forest Service had in British Columbia. He pointed out that this discrepancy means the public has a choice between giving up public ownership of the forest and getting over its dislike of large bureaucracies, since the 3,000 people on staff were woefully inadequate for the task they had to perform. But that took Pearse two and a half pages to say.

Because he loved too much he wrote too long, so the public missed lines like, "A striking feature of the present allowable cut calculation is its fine analysis of crude data which produces spurious precision in its results."

That single sentence captures the situation. To try to do the same for the Pearse report: this most thorough study of the government and the industry failed because it was so exhaustive.

It takes the fury of an angry god, the showmanship of P.T. Barnum, and the guile of a pickpocket to get the public excited enough to pressure the government. Instead of these, Pearse used the skills of an analyst and a researcher. And that left the government free to pick and choose what it wanted from his report.

Chapter 25
Intermission: Meet the Cast

A lot of people flunked out of the north. A lot of people still do.

One pulp company official said, "We find that a good number of management people we hire leave after their first winter. Either they can't stand it, or their wives insist on going."

That's because, with the exception of the bars, the only entertainment in most towns is what the people create for themselves. Each fall, across the north, hundreds of clubs go into action, catering to hobbies, social interests, and dozens of other common bases for getting together. Often people are members of several different organizations.

In contrast to large cities, in the north you can't just pay to be entertained.

A lot of people hate it.

Those who stay get changed by living with themselves. A few years in the north and you either have less need to be entertained or else you go "buggy."

One of the major sources of entertainment is the outdoors. No one has ever moved to Vanderhoof for its fine restaurants. No one has moved to McBride to see foreign movies. Here are some snapshots of life in the north, during the boom, and today.

1. *Man riding stuffed moose*

That's the old MacDonald Hotel. Place where you could pick up a hooker, drugs, or a knife in your ribs. Burned down several years ago. The rebuilt hotel has far fewer knifings than it used to, but it's still an educational place to take visitors, especially if they've led sheltered lives. For anyone convicted of a crime involving drugs or violence in Prince George, a common condition of parole may be staying out of a six-square-block area downtown that includes the Mac.

Of course the Mac isn't the liveliest bar in the north, nor the one with the worst reputation: that's the "Zoo" in Fort St. James, the place with the sign over the bar warning that "patrons must check hunting knives before being served." During the mid-1970s an ex-New York street fighter visited the Zoo and left terrified—at 3:00 p.m. on a Thursday.

2. *Photo of the bride—pregnant*

In a lot of lumber towns in the Northern Interior, including Prince George, Mackenzie, and Houston, about half the population over nineteen is younger than thirty-five. This segment of the population has had some of the highest income levels in Canada. It's been uneducated, with millworkers averaging grade eight in 1966, grade nine in 1978. In Prince George seventy-eight percent of the population over fifteen years old is now or has been married. In 1978 there were 14,645 families in town, and the birth rate during 1975-78 averaged 1,478 per year. A shade under ten percent per year.

Until recently, a guy could quit school at seventeen and be sure he could get a job in a mill, making enough money to buy a Trans-Am and get his own apartment. This easy money for drop-outs encouraged dropping out, according to dozens of teachers in the north. "When the kid says, 'Why should I bother? I can get a job in the mill,' what can I say? In five years he'll be making more money than I do."

Often within a few weeks after he buys his Trans-Am, his girlfriend announces she is going to have a baby. Teenage pregnancies are not unusual—they're far more common than first pregnancies among women over twenty-three.

It can be argued that since there isn't a single mill in the province with minimum education requirements, the industry's hiring practices encourage pregnancies. But it's a lot more realistic just to say the mills encourage dropping out of school.

3. *The Prince George sports car*

The three-quarter-ton Ford pickup with four-wheel drive, chrome-plated chain on the winch hanging from the front bumper, twin whip antennas, and a bug deflector with the owner's Citizen's Band "handle" on it may belong to a millworker, or to a doctor. You can't tell from the canoe in the summer, the snowmobile in winter, or the "Easy Rider" gun rack in the cab. In the north, trucks like this one seem more common than compact cars.

4. *Orchestra versus skin magazine*

The Prince George Symphony, started in 1970, performs to sold-out houses several times each year and has a staunch following for all performances, helping to make it one of the few orchestras in all of Canada that has managed to remain in the black. In 1979 it was the only orchestra in North America to perform the difficult piano concerto No. 4 by Rubinstein. Many of the players have performed with leading symphonies.

But the largest distributor of magazine and paperbacks in the north, Gundy's Magazine Service, reports that *Penthouse* outsold every other magazine that year, with *Model Aircraft News* coming in second. *Cosmopolitan* is the best-selling women's magazine. Some grocery stores sell more than 400 copies of tabloids like *The National Enquirer* and *The Star* each week.

If you want to know where to stay anywhere in Europe, there's someone in almost every lumber town who has just returned from England, someone else from Germany, someone else from Katmandu. If this sounds exaggerated, please tell that to the teacher who earnestly tried to convince a polite high school student that traveling to Europe would broaden her world view—only to discover that the student had spent a year in Paris. Or tell that to the doctor who just returned from a short vacation in Africa to discover that one of his patients had worked there for five years.

And, of course, it is from Prince George that the Garden of Eden, a mail order lingerie and personal electric appliance supplier, gets most of its business.

The north is a contradiction. It has its own culture, and to look at any single aspect is misleading. But if there are bonds, if there are things that distinguish those who stay from those who leave, they're in an attitude and an opportunity.

Those who survive in the north, and who want to stay in the north, have learned how to entertain themselves. Book stores are rare. Record stores have limited selection. A lot of buying is done by mail order. The theater is limited. Restaurants of high quality are far rarer than in major cities. People who are to survive the long winter have to be able to entertain themselves through hobbies, or through joining (or starting) an organization that does things.

This leads to a lot of social activities in the winter months and, among many, an almost desperate urge to get out into the outdoors whenever the weather is good. In the north, there is the opportunity found in small towns throughout British Columbia that is missing in Vancouver and Victoria. It's the ability to get into the forest in just a few minutes. The excuse may be fishing, hunting, canoeing, backpacking, or picking raspberries for jam, but the "reason" is often merely an excuse to get away.

When it's possible to have a lake to yourself, half an hour from home, six people make a lake feel crowded. When it's easy to see wildlife (*everyone* in the north has bear stories), watching a marten on a tree branch becomes an expected, but always new, pleasure. The quiet of fly-fishing, with no one around to comment on how the line hits the water like a falling tree, the smell of fall, and the other pleasures of solitude often become a way of life.

Recreationists at Finlay Junction, 1914.
Photo courtesy B.C. Forest Service.

Chapter 26
The High Rollers

In good years, sawmill owners and wholesalers often gamble. They'll load a railcar with lumber and ship it east to Minnesota—but without having any idea who's going to buy it. If a buyer is found while the carload is en route, a buyer who needs quick delivery and is willing to pay a premium, the millowner wins. But if his "roller" rolls east and ends up sitting in a freight yard, the millowner loses, since he'll have to pay for yard space and car rental. There are no figures on how many rollers go east each year, since firms want to appear more respectable than riverboat gamblers. But as Van Scoffield, manager of the Council of Forest Industries' northern office, said in 1978, "Statistics show more B.C. lumber is shipped to Minnesota than to anywhere else in the United States, and we know Minnesota doesn't use *that* much wood."

Another gamble: lumber futures are traded just like hog futures and wheat futures on the Chicago Mercantile Exchange. The way it works is simple: all you have to do is grab a crystal ball and guess the future. A mill may calculate that the future is going to be terrible and so will jump at a contract to produce two-by-fours at $166 per thousand feet, with delivery in eight months. But if the price of lumber starts to rise, the mill may say, "I'm going to be stuck getting a rotten $166 per thousand," and it will buy the contract back, perhaps for $172. If the market only goes up to $168 by delivery date, the mill loses. If the market goes up to $178, the mill wins. As in all commodity markets, more people who buy futures lose than win—with the losses acting as a cushion for millowners in bad times.

Not long ago, there was really only one place to gamble: the U.S. housing market. Every mill in the north turned out lumber for houses or rough lumber that was finished and shipped to home builders. Well into the 1960s, the "rail market," which meant North America, took almost all of the output from the north. In the 1970s the amount of lumber being shipped by northern mills to Europe, the Pacific Rim, and Australia began to expand, but until the end of the decade most lumber went by CN, CP, or BCR to the buyer.

With everyone competing for the same wholesale buyers in the United States and producing the same products, firms quickly learned how to hedge their bets. They proved that all two-by-fours are not created equal.

If you cut a board from the very heart of a log, the grain pattern will look like parentheses: ((())). The board won't warp. Cut from the outer edge of a big log, the grain could end up looking like a bunch of U's stacked on top of each other. Wet the second board, and it will curve like a teacup. Even if it's straight after being dried under pressure, being left on a building site may make a supposedly straight board fit only for the railing on a circular staircase. That's why some mills in the north, like Lakeland, that produce two-by-fours and almost nothing but two-by-fours get a higher price per stick than mills that use the center of the log for large timbers and use the leftovers to make two-by-fours.

There are other ways to make two-by-fours different from other two-by-fours. Mills that supply the do-it-yourself market put an extra fine finish on their boards. The owner of a home in Oak Park, Illinois, who commutes to an office in Chicago probably knows nothing about grain, species, drying, and the technical tables in the back of engineers' handbooks. But he knows what looks good and buys the smoothest, best-looking board he can.

On the other hand, a housebuilder in Ayer, Massachusetts, isn't as finicky about finish on lumber that's going to end up hidden inside the walls of a home. But he insists on the proper mix of grades in every bundle he buys so that there will be good wood for structural purposes and he can avoid paying a premium for the bits and pieces that end up being used for things like the second two-by-four under a window opening.

Other firms find other specialties. Some firms supply stress-rated material used in the manufacture of prefab trusses—the triangles of boards that hold up the roof of a house.

But there are only a certain number of visible differences a mill can provide, like painting the ends of its eight-foot-long studs blue, red, yellow, or green. And there are only a certain number of finicky buyers. So some firms stress invisible differences: quick, accurate billing, or quality control. One lumber seller said, "Deliver one load of bad wood and a customer's gone forever." According to another one, however: "If your product is 50 cents under the competition, everyone will buy it. If it's 50 cents over, no one will buy."

To get around this head-to-head competition, some firms deliberately avoid making what other people make. Jim Rustad, of Rustad Bros. in Prince George, said:

> *We don't even make two-by-fours. We customize each order for every one of the hundreds of wholesalers we serve, and unlike some companies we make special sizes to fit a customer's needs. Every three weeks, for example, we ship a carload with two special sizes to a company in the Southern States that makes walk-in coolers.*

Constantly resetting saws and customizing may seem like a lot of work. But an extra dime per thousand feet of lumber means a lot to a mill that makes 100 million board feet of lumber per year, as many do, and to some firms an extra buck means $0.25 million to $0.5 million.

The mill manager obviously doesn't have time to call around the world to see who wants what and how much they're willing to pay. So large firms have lumber traders on staff. Smaller firms sell either through large firms or through organizations like Seaboard Lumber or Eacom, or to independent lumber brokers. The brokerage organization usually takes about a two percent commission for its efforts in matchmaking between buyers and sellers. But they gamble too. Often wholesalers will contract with mills for products they think they can sell at a certain price. Sometimes they win; sometimes they lose.

In October 1973, Don McKay and some friends started a lumber trading firm called Skagit Industries. The first year they broke even. "The second year the market was bad and we lost $50,000," Don said. "The year ending

October 31, 1976, we made the $50,000 which offset our position. The following year we made $250,000; the market had come back up. The year after that we made $300,000. I considered it to be making money."

And that was a *two percent* commission. In other words, the firm handled about $15 million in lumber, and it was one of the smaller trading firms in the province.

This type of lumber selling, with its multitude of buyers and sellers, is "just like Economics 101 with supply and demand determining prices," according to Bill Johnson, assistant to the vice-president for wood products marketing at Boise Cascade in the United States. And the vast majority of wood is sold this way. But each year a growing amount is handled away from the commodity market.

Rail lines that used to serve small towns have been abandoned. Lumber cars have grown in size. Instead of shipping mixed carloads to local lumberyards, most lumber is shipped to regional wholesale distribution yards, which send mixed truckloads to retailers. Each year the number of these yards has grown, until several firms control over 100 each. "We sell more than we manufacture," Joe Bennett of Georgia-Pacific said. "That lets us run our mills at the rate we prefer."

The idea appeals to B.C. mills. Some have distribution yards. Others, like West Fraser and Crown Zellerbach, own retail lumberyards. Other firms have long-term contracts to deliver a certain number of carloads per month to wholesale yards for years to come.

In the mid-1970s these "tied sales" amounted to only a few percent of the lumber production in the north. And mills were sick of riding the roller coaster of U.S. housing starts—especially since the roller coaster went downhill so regularly. More and more firms began exporting overseas. The Council of Forest Industries' Northern Interior Lumber Sector pressured the Council to do the missionary work needed to get Japanese and English builders to use two-by-fours in their buildings. They were partially successful. They got builders to use lumber for the structure, but in their own sizes, like $1\frac{7}{8}$ x 7".

Later in the decade the mills discovered a new gamble that made "rollers" look safe by comparison: shipping to the Middle East. In Kuwait, Egypt, and other countries in the neighborhood, a lot of importers bring in television sets, clothing, and lumber. They don't understand Canadian grading standards (unlike the importers who specialize in lumber in these countries). All they know is price—until the lumber arrives. One firm assembled an entire shipload of lumber, sent it to the Middle East, and watched the buyer tear up the letter of credit for payment right on the dock. The firm went bankrupt. Another firm called a trade magazine reporter and said excitedly, "We shipped to the Middle East and got paid for it!"

No firm wants to be thought of as a gambler, but the forest industry has more high rollers than ever appeared in "Guys and Dolls."

Chip N'Saw. That isn't waste on the floor, that's money pulp mills will pay. Photo courtesy of Canadian Car.

Chapter 27
The New Forest Act: 1978

On December 11, 1975, the NDP was defeated. Social Credit was elected government. Tom Waterland, a mining engineer, was named Minister of Mines, Forests and Water Resources.

Every reporter in the province immediately played the politician game. Catch a freshman politician young and ask, as soon as he's made a cabinet minister, "What are your plans? What are you going to change? What do you like about the forest industry? What bothers you about the forest industry?"

Some politicians are dumb enough to answer, before they have time to find out, from the inside, what's going on.

Tom Waterland—with hair that looks like a blond brillo pad, glasses that give his face a kindly appearance, and a slow-talking friendly manner—didn't. He said he had to learn about the forests and the forest industry.

This gave him a breathing spell. He could go into meetings with the Forest Service, industry, unions, environmental groups, or anyone else and say, "What do you think I should do?" Then he'd listen intently, ask questions, and at the end of the meeting the people he talked with felt that they'd talked *with* the Minister, not *at* him.

Put yourself, for a moment, in his shoes. The unions wanted to make sure that there'd be employment year round for all the members. The industry was almost willing to kill for security of timber supply, since quota, "third band" timber, and most of the rules of the game weren't covered by any legislation. And the NDP's time in power had scared them into realizing that, in law, their tenures were tenuous. The Forest Service had its own problems. The most recent legislation giving direction to the Service had been back when sustained yield was brought in. Money was spent according to the area of interest of the Forest Service official receiving it. There wasn't enough money to do an effective inventory, or for the replanting needed for sustained yield, or for any of a dozen other things the Forest Service was supposed to do. Environmental groups felt, rightly, that the legislation in force was largely logging law, not forestry law. And every other Ministry wanted more money for its own projects.

That was bad, but the options open to the government were worse. Waterland couldn't wipe out quota, even though legally it didn't exist. People had bank loans based on it. Mill construction had been based on the expectation of a continual cut of a certain size. The economics of the industry depended on quota to the point that, when asked how much the industry would spend to get rid of a government that wiped out quota, one executive replied: "Oh, perhaps sixty million dollars a year."

"SIXTY MILLION DOLLARS! Why that number?"

"Anything else would be overkill," he laughed. "The point is this: the industry will spend whatever is necessary to retain the present quota system."

The industry saw competition for timber rights as a threat. Because there was no history of competition, bringing it in would be very difficult. People fear change; and since change might mean added costs—and that would mean lower profits—it would be in the industry's best interests to oppose change, so that up to seven tons of logging slash per acre could continue to be left behind after trees were harvested, and so that logs twenty-one inches at the butt could continue to be fed into a portable chipper on Vancouver Island.

If the government ignored the industry and withdrew land for parks and ecological reserves, a small percentage of the people in British Columbia would be delighted. But the woods unions would march on Victoria since withdrawals of forest land for purposes other than lumber threaten jobs.

Many opinion polls have been commissioned to see just what the government can and cannot get away with. One taken during this period, by a respected organization, said the government could impose more control over the industry without getting the public upset and could take a bigger slice of profits. But it had to be done very carefully.

Tom Waterland acted, quietly, within the range of "Ministerial discretion" in the old Forest Act. Throughout his career Waterland followed three political principles that are sound no matter what party a politician belongs to:

1. Misdirect attention so people will look at what you want them to look at.

2. Always have something else due to happen, something else in the works. Protests from people who would otherwise oppose the changes you are making will be muted for fear of "the next shoe you have to drop."

3. Never tell what your ultimate goals are, what you are trying to accomplish as a whole. If Tom Waterland had told the forest industry all the changes he'd make, resistance would have been well financed and well organized. As one envious NDP politician said, "Tom Waterland has gotten away with far more than Bob Williams ever dreamed of."

In almost a dozen White Papers, lots of committees, lots of changes involving a new Forest Act, a new five-year program for reforestation, and a reorganization of the Forest Service, among other things, Waterland has been extremely successful at having more shoes to drop than a centipede.

While the public and the industry were watching these things, the Minister approved licence transfers that led to every region of the province, except Williams Lake, having one or two dominant firms. Often there were other bidders who might have provided a better balance.

These transfers of rights involved far larger firms than Ray Williston had talked about when he spoke of weeding out the firms too small to compete. The cutting rights involved often ran to the equivalent of over 100 million board feet per year.

The Pearse Commission report didn't come out until September 1976, almost nine months after Waterland took office. In the meantime he could say things like:

> *I'm basically against government involvement. When government gets involved, it usually doesn't do it very well.... There's a tendency for any administration, private or public, to grow like topsy, and you end up with a structure that doesn't fit what you're trying to do. (But) people shouldn't have to wait six months for permission to build a road they'll have to haul logs on next year....*

When the Pearse Report was at last released, the Minister wisely said he'd be a fool to comment on it until he had studied it thoroughly. Then he appointed a select committee, the Forest Policy Advisory Committee, headed by Bob Wood, a consultant and sawmill owner, to come up with a new Forest Act.

Before the new Act was introduced, Waterland said that while he was sure tenure would be discussed in the Act, the amount of timber a company could cut would depend on a Forest and Range Resource Analysis, and, to get the best possible data, inventory activities of the Forest Service would be given more emphasis—and money. So before the Act came out, the industry knew that the new Forest Act wasn't the last word on the subject. And when the bill was introduced, it often spoke in generalities, with the details to be given in yet-to-be-announced regulations. Shoes kept dropping, with another one always to come—delaying the day when the industry would have time to evaluate Waterland's changes as a complete package. As each shoe was added to the pile, the options open to the industry were removed, little by little.

One result was that the industry always had something to yell about and yelled so often that the public stopped hearing what it was saying (if anyone had listened in the first place). The government soon had so many White Papers and studies, committees and task forces, that the industry was busy reacting to what the government was doing rather than pushing the government. And everyone was so busy watching the dog and pony show, complete with everything but dancing girls, that Tom Waterland was able to continue to do what he wanted, often using discretionary powers for decisions that never reached the limelight.

After the sixth White Paper, Waterland was asked when he'd be willing to rest on his laurels. "Never," he replied with a smile. According to one senior official, however:

> *If you look at what's been introduced, what's being introduced, and what we've said we're going to introduce, you'll notice that while we're still dealing with important issues, the framework has already been put into place: the new Forest Act, a reorganization of the Forest Service, and a plan for action, with goals. But the industry has been conditioned by now so they respond to everything with equal intensity. They haven't noticed that most of the big issues have been settled.*

That's an exaggeration. The Forest Act introduced in 1978 was modified in 1980. The Small Business Program—suitable for the logger with six strong sons—was still causing problems in August 1981. But the basic idea, that the industry hadn't noticed how many changes it had already accepted, was true.

In the fall of 1978 Tom Waterland and Bob Wood were interviewed. Bob Wood had been Bob Williams's executive assistant, he said, and he'd offered his resignation when the government changed.

> Bob Wood: *Tom said, "Let's see how it goes for six months" to bridge the transition, and then I went to work with FPAC (The Forest Policy Advisory Committee) on the new bill.*

> Tom Waterland: *We had developed a good way of administering the forest. It required some streamlining and some legal form. I don't want to get into percentages, but we also tried to overcome its shortcomings.*

Bob Wood: *It was government by fiat. The old bill said "the Minister may" and the rest was commentary. But that's a pretty shaky way of doing business.*

Tom Waterland: *And we've provided incentives in the Act as distinct from—*

Bob Wood: *Now the Minister may do very little. We've tried to get politics out of the forest. We've provided stability of raw materials and there's flexibility in the Act based on performance.*

Question: *Half of the new Act seems straight out of Pearse's report.*

Tom Waterland: *Let's not get into percentages. For eighteen months we kept asking ourselves, "Was Pearse right?" He's an academic and he took an academic approach to the problems. We couldn't take any of his ideas and put it straight into the Act.*

Question: *But the concepts he used—*

Tom Waterland: *More than fifty percent of Pearse's recommendations appear in some form or another in the new Act.*

Question: *The new Act doesn't really do anything about competition.*

Bob Wood: *The pie is carved. The competitive elements have shifted from getting more woodlands to intensive forestry and utilization....*

Soon Tom Waterland had to leave. The conversation with Bob Wood continued.

Bob Wood: *The dilemma is public land ownership.... It's political nonsense to suggest selling public land, but that means you've got to acknowledge it's the Crown's and that it's up to the Crown to manage it. By definition that means creating rules that are "non-free-enterprise."*

Question: *Which only a free enterprise government could get away with. If Williams had tried it—*

Bob: *That's right. And the basic motive of the Crown has to be enhancing the resource, while the companies keep their costs competitive, or they're dead.*

The new Forest Act could only have been introduced when it was introduced, and there's been speculation that the government waited for a strong lumber market before bringing it in. In 1976 the selling price of forest products had slowly begun to recover. Plateau Mills' new 150,000 board feet per shift sawmill started in August 1976, complete with computer-controlled log sorting. Carrier Lumber in Prince George had a new mill that hit 250,000 board feet per shift within three months of start-up that year. In September a new sawmill was announced for Houston to be jointly operated by Eurocan and Weldwood. Lumber sorters began to be installed at many mills, replacing the old green chain—the guys who stood waiting for lumber graded to one or two standards of a single size and then stacked it as it rolled down a conveyer. A mechanized lumber sorter could handle over sixty different sizes and grades, with just a few men keeping watch. This usually cut the sawmill staff by at least fifteen men per shift.

In 1977 the market improved a little more and West Fraser put in a $7 million sawmill at Smithers to replace the old Pacific Inland Sawmills and Fink Sawmills plants, which they'd bought. At Houston the Weldwood-Eurocan joint venture, Babine Forest Products, was underway. Apollo Forest Products in Fort St. James completed its new mill, and Northwood opened its new headquarters—one of a handful in British Columbia actually built of B.C. lumber. Soon other firms, like Rustad Bros., Lignum, Gregory Industries, and Tahsis, built offices that featured their own products. Lakeland Mills acquired Helco Forest Industries and began planning a new mill.

The next year, 1978, was even better. It was the best year the forest industry had ever had. In June Tom Waterland introduced the new Forest Act, which at last put "quota" into legal form. The industry was delighted, and the NDP called it a sell-out. Few people noticed that although the Act restricted "Ministerial discretion," there were enough loopholes that the Minister could still do what he wanted.

Today people in the industry, at universities, and in the government study the Act with the detailed care usually reserved for scripture. Here are a few of the main points:

Public Sustained Yield Units were to be combined with all other forms of tenure and all other provincial forests to make up "Timber Supply Areas" (TSAs). Instead of granting cutting rights in a particular location, the Crown only guaranteed a licence holder an Allowable Annual Cut within a far larger area. The idea was that instead of arbitrary marks on a map, each TSA would be naturally bounded, usually containing entire watersheds, so that mills in a given area would find all their cut within one, or at most two, TSAs, and all within a hauling distance that made economic sense.

And of course lumping several of the old units together hid management mistakes in some PSYU's.

New types of tenure were announced. Tree Farm Licences remained almost the same, but the major form of cutting right was to be a "Forest Licence," with automatic renewal every five years, containing cutting rights to approximately the same amount of timber the firms had held under most other forms of quota.

Pulpwood Harvesting Agreements continued. Mill Licences could be required, so a firm couldn't expand first, then say, "Either we get more wood or we lay off our staff."

The Act promised that each firm's cut would be maintained within five percent each time a Forest was renewed. If the reduction in cutting rights was more than five percent, the Crown had to actually pay companies for their loss of cutting rights. But the Act had a simple escape clause: all the government had to do was cut down the size of the Allowable Annual Cut for everyone in a Timber Supply Area by equal percentages, and it didn't have to pay.

An appeal board was set up so that people who disagreed with the regional manager, Chief Forester, or other Forest Service official could go to an independent board in many cases. But after the Forest Service lost the first appeals under this part of the Act, the rules were amended, cutting down the number of things that could be brought before the board.

The amount of advertising required when the Crown decided to give out timber was cut down in some categories.

There was a promise that if firms improved the yield of timber, their annual cut would be increased, and that they could get money from the Forest

Service for planting, fertilizing, nurseries, and so on.

The Timber Products Stabilization Act was repealed, and the Logging Tax Act was amended almost out of existence.

It *sounds* as though the companies got everything they wanted. They did get tenure and quota lodged in the Forest Act. But experience would demonstrate that the Forest Service and the government had actually legislated firmer control over the industry, with most of the power in regulations that could be quickly changed. The Act emphasizes plans and planning before logging, for example. All the Forest Service would have to do is keep saying, "Your plan is unacceptable," to a company, then use another section of the Act to say, "You didn't cut enough of your Allowable Annual Cut, so you lose what you didn't cut."

Detailing the many ways the government and/or the Forest Service can help or hurt a firm would take several pages and would include things that have never been done, as well as things that are common. Here are a few examples of the types of control the government has:

If a firm owes money for stumpage, the government can take debtors to court—and it has done so. Or the government can, and has, let firms avoid payment until the market goes up.

A firm can pay for an area to be cruised, decide that it's an area that is worth logging, ask the Forest Service to hold an auction, and be told, "There isn't enough merchantable timber in the area to justify a sale." Then the government can let someone else put the area up for auction and give it to the second firm. This has happened.

In theory, but not in practice, the Forest Service has the right to close a firm's logging operations from January 1 to December 31, allegedly because of "fire hazard."

The regulations provide for a certain amount of debris as normal to logging operations. Then firms have to pay for additional levels left behind. Although the Forest Service has often been generous in its assessment of the amount of debris that may be left, a slight change in regulations (or merely a close examination of the debris left by some companies, especially on the Coast) would raise costs, lowering profits.

The new Forest Act lets any government do what it wants to the forest industry as a whole, or to individual firms, without changing a word of the law.

But the firms, in their joy at finally getting the alienation of the majority of Crown forests into law on a basis that provided perpetual renewal as long as they didn't mess up too badly, didn't notice this.

It's fitting that when Tom Waterland spoke about the new Act to the legislature on June 14, 1978, he said:

> *The 1978 Forest Act replaces a very old and somewhat creaky, and, in many aspects, obsolete 1912 Forest Act....*
>
> *Mr. Speaker, neither the royal commissioner nor this government proposes drastic revisions of the existing system. Some self-interest groups have clamoured for wholesale redistribution of timber rights. Again, neither this government nor the royal commissioner reccommended this, nor is there any valid reason to do so....*
>
> *I would like to quote the remarks of the Hon. William R. Ross, Minister of Lands, made in this Legislature in 1912, when debate began on the second reading of the 1912 Forest Act.*

"Sir, a certain solemnity of its own surrounds the introduction of this forest bill. An epic is condemned, a new epic inaugurated. A turning point in the development of this great, young province has been reached. We raise ourselves today above transitory interests of this week, this year. We glance down the vista of the years to come, and, turning from that vision of the future, we call the world to witness that we legislate today not only for ourselves and the needs of this day and this generation, but also and no less for our children's children and for all posterity, that we may hand down to them their vast heritage of forest wealth, unexhausted and unimpaired."

Prince George, 1950. Photo Courtesy of IWA local 1-424.

Chapter 28
Secrets

For a long time forest company presidents didn't have to deal with, or care about, the public. Their only concerns were unions and the government. As a result, some company presidents squirm and try to get out of telling anything about their firms. Others react more strongly. Their logic is, "If the public doesn't know what's going on, they won't get upset—which they might because of a lack of understanding of our business." That may sound like a contradiction, because it is. However, several companies in the forest industry, including some in the Northern Interior, follow this rule. As the president of one major Interior firm said, "The forest industry acts like it has a guilty conscience."

That's one reason people don't know what's really being done with the public forests.

Another reason is that provincial governments—no matter what party has been in power—have been less than candid. After all, the government has bungled its management of the forest. Trees, like coal, are a diminishing resource. No government has succeeded in putting the province on a truly sustained yield basis, replacing the trees harvested each year, although it's allegedly been trying to do so since 1947. The government has made the forest industry what it is today—good and bad—but if the voters realized this, they might disagree with what has been done and what the results are. So governments tend to sing songs of praise about some of their actions and shut up about others.

Unions are more open than the government or industry, but calling the IWA to ask about the way personal income tax dollars subsidize union settlements leads to harangues filled with convoluted logic.

The Forest Service is, in some ways, the worst of the lot. When research for this book began, and documents were requested from the Forest Service, the official policy was that only those documents that had been tabled in the legislature were officially public records. When asked about contracts made forty years ago, the Forest Service replied that they were confidential. When asked if old records could be gone through, the official response was that a list of questions should be supplied. The questions would be examined to see if the answers were suitable for distribution to the public, and then the records would be gone through by a staff member, at a cost to the inquirer of $15 an hour.

Luckily, many members of the Forest Service thought this policy was stupid. Bill Young, the province's Chief Forester, has now changed the policy, explaining:

For years it was a rare thing for anyone in the Forest Service to even be approached by a reporter or someone off the street. And when forestry started becoming an issue, a lot of people didn't know how to handle it.

Thanks to Bill Young, Mike Apsey—the deputy Minister of Forests—and Tom Waterland, more information is available than ever before. However, contracts, the amount of stumpage paid by individual firms in any year, and the names of the firms that owe the government money are still confidential.

The reason everyone concerned with the province's forests is so secretive is simple: they all have things to hide.

Here are a few of them.

THE INDUSTRY

The two things the industry doesn't talk about are "lumber recovery factors"—how much of a log is actually turned into lumber—and profits. Trying to get honest lumber recovery factors from enough firms to figure out the average, and who's wasting wood, could defy the talents of the CIA.

Profits are another matter. Although the industry constantly complains that it's about to be driven over the hill to the poorhouse, the only way it will be driven is in a Rolls Royce.

The owners of reasonably efficient sawmills often average sixteen to twenty percent return on investment per year. However, this isn't all they get. Les Reed, who's responsible for forestry for the Federal Government, said, "Anyone who forgets that the big profits in the forest industry come when you sell out is a fool." Although the price paid for timber rights is written off by some firms within a handful of years, most of the price paid for forest companies is for timber rights. And timber values have been steadily increasing for years.

When the market goes down, sawmill owners often cry doom and gloom. But in 1981 the typical price to produce a thousand feet of two-by-fours eight feet long—the standard studs used in home construction—ran about $148. Some firms had costs of $156 or more. Others ran under $142. Although these numbers don't include interest the firms have to pay on long-term debts, or the cost of new equipment, they do suggest how low prices have to dip before the majority of firms suffer. Companies were complaining when prices dropped to $172—a measure of sound versus suffering.

Some real numbers: in 1978 West Fraser had $16 million in certificates of deposit. Two months later in that good market year the company had $20 million. In the north, during the years 1977-79, companies that owned both pulp and lumber operations made about $50 million a year. This figure didn't necessarily show up on the annual reports, since some of the profits went to buy out other firms, pay for new equipment, and pay off long-term debts.

Another measure of suffering is how many firms go bankrupt. Three-quarters of all retail businesses go bankrupt within two years of starting up. But in the last half dozen years, only a handful of forest companies producing more than 10 million board feet per year have gone bankrupt. Some have sold out for less than the owners would have liked, but as long as cutting rights have been held, the owners of even the worst-run firms have been able to walk away with money in their pockets.

This isn't the way the system is supposed to work. According to the economic theories taught at universities, the firms with the best managers and machinery should outperform the ones with mental midgets in charge and junk equipment. The bad firms should be driven out of business. Because of the government control of the forest industry, however, this doesn't happen.

THE UNIONS

The amount of money mills pay the government for the trees they cut fluctuates. In poor markets the stumpage rate goes down. But it also goes down when the cost to log a stand of trees goes up. So every time a logger or millworker gets a dollar raise, the stumpage rate under good market conditions goes down a dollar.

This subsidy, which covers industry wages, means that wage increases of a dollar really cost the forest industry about 70-85 cents. This makes it easier for the IWA to get reasonable—from the union point of view—settlements than for unions in industries that do not have a subsidy. This subsidy has three effects:

1. Since the subsidy costs the government money, each taxpayer in the province contributes to IWA wages in the form of taxes that are higher than they'd otherwise be.

2. The government has been subsidizing the union members and industry for so long that changing the system would be very difficult. Management and the union would band together to dump any government that tried to derail the gravy train.

3. According to a discussion paper prepared for the Economic Council of Canada—but never released because of pressure from the forest industry and unions—called "Natural Resources and Regional Disparities: A Skeptical View," by Lawrence W. Coptithorne, this subsidy causes strikes in other industries. The IWA gets a dollar and people in other industries say, "They got a dollar, we should get a dollar." But in those other industries, there's no subsidy. So management is more reluctant to give a dollar and offers 85 cents, for example. Here comes the strike.

THE GOVERNMENT

Tom Waterland said, "Since the turn of the century governments have consciously used the forest as a tool for economic development." And as a source of revenue. In the fiscal year ending March 31, 1980, the government took in about $361 million more in revenue from the forest than it spent. But this figure includes only direct revenue and direct expenditures. Add things like the province's share of corporate income taxes, and the government's own figures show that the province re-invests less than a quarter of the money it receives.

That's one reason why the government has never managed to put the province on a sustained yield basis, replacing the trees that are cut each year. Instead there are about 1.8 million acres of harvested land that have grown back in brush, or where trees are packed so closely together they stunt each other's growth. The government is doing almost nothing about this Not Satisfactorily Restocked land, and it never has.

Of course, saying the government should act is easier then footing the bill. According to Ray Williston:

> *When you're in the cabinet for forestry, you're up against education and health. If there's money in the kitty, they get it and you get the leftovers.... I used to curse whenever the sliding scale (on stumpage) went up. The Treasury Branch would expect the same revenue the following year. My life would be hell warmed over whenever the market dropped.*

But where is the government going to get the money needed for a proper reforestation job? Should it close the hospitals, the highways, or the schools?

The government has other things it really prefers not to talk loudly about. All governments in the past decades have said, often, that they believe there should be a diversity of sizes of firms in the forest industry. Meanwhile, companies keep disappearing. In Peter Pearse's report *Timber Rights and Forest Policy* there's a chart showing the thirty biggest firms in the province, firms with good timber supplies, sources of capital, and mills big enough to be—allegedly—efficient. Since the report was published in 1976, five of the thirty have been taken over by other firms. The more than fifty takeovers in the forest industry in the past nine years have often involved direct results of forest policy.

Some effects of government policy aren't what the government intended at all. Companies pay the Workers' Compensation Board rates based on the number and types of claims that are made to the board. So in at least one pulp mill in the north, workers are told they shouldn't file WCB forms when they're subject to chlorine gas as the result of a leak in the mill. This advice leaves the worker without recourse if a lung collapses a year or a decade later. When John Asling and Eli Sopow, then reporters for the Prince George *Citizen*, researched an article on chlorine gas leaks at pulp mills and the damage they did, one of the mills—Northwood—had both union and management discuss the problems at a joint meeting and talk about what was being done to solve them. At another mill, management dragged its feet. At one interview the reporters were told, "It's no worse than the chlorine in your drinking water." "Hey, I worked in a pulp mill," Sopow replied. The company changed its tune. Asling and Sopow were asked by the newspaper to rewrite their story "to make it more positive," then to rewrite it again, and again. It never appeared in Prince George, although CBC radio in Prince Rupert did carry the story.

The government never intended this result. But government policy does not enter the world as imagined in Victoria when the legislature debates an issue. The real world is one of less-than-perfect people, less-than-perfect policies, and all-too-perplexing problems.

THE FOREST SERVICE

When Bob Williams had the Forest Service do some logging in an attempt to get realistic cost figures, one logging firm owner was delighted. "It took them ten men and $10,000 to match the efficiency of the average beaver," he said.

Until just a few years ago the Forest Service didn't have the right to examine a company's books to figure out real costs so stumpage rates would reflect them. "In the old days, we took whatever data we could get, were grateful for it, and averaged it out," one official said.

The Forest Service has never been given the staff, money, or regulations needed to do the job the law called for. Between 1947 and 1978 that job was allegedly to get the province on a sustained yield basis. But without support from the government—meaning money—it was impossible.

Forest Service members who complain in such a way as to get the government of the day in trouble end up in trouble. "If you embarrass the government or the Forest Service," one ranger said, "they send you to Atlin or Victoria—some place safe." Another warned, "Use my name in print and I'll end up in Alice Arm."

The Forest Service has always tried to give the impression that the situation was under control, even when it wasn't. For many years, it reacted with paranoia whenever it was criticized by anyone. Only now is the Service beginning to lift its veil of secrecy a little.

It's obvious that there are a lot of things that could be improved: by the Forest Service, by the government, by the unions, and by the industry. The question is how to improve them. You've read how the industry evolved along with government policy and how the system really works. What happened in the past, causing present attitudes and rules, can't be ignored. But none of the groups involved in what happens to British Columbia's forests, including the voters, are prisoners of history.

Bibliography

A bibliography does four jobs. It credits books, articles, speeches, and papers quoted. It acts as a guide if you want to find out more about the things discussed in the book. It answers the questions "Oh, yeah?" and "Says who?" And it lets the expert reader send a letter mentioning something the author didn't read, but should have. On the other hand, mentioning the books read, but unused, merely sends others up the same blind alleys. All of the sources listed provided facts, quotes, or ideas in *Cutting Up the North*.

Angier, Bradford. *How to Live in the Woods on Pennies a Day*. Harrisburg, Pa.: Stackpole Books, 1971.

Barron's (magazine), April 23, 1973.

Bernsohn, Kenneth. *Slabs, Scabs & Skidders: A History of the I.W.A. in the Central Interior*. Prince George: IWA Local 1-424, 1978.

────────── "Lumber Marketing." *ForesTalk*, Summer 1980.

Black, E.R. "British Columbia: The Politics of Exploitation." In *Exploiting Our Economic Potential: Public Policy and the British Columbia Economy*, edited by R.A. Shearer. Toronto: Holt, Rinehart and Winston of Canada, 1968.

Board of Trade of Prince George. "Prince George British Columbia." (pamphlet) 1919.

B.C. Magazine, October 1911.

British Columbia Provincial Government:

Forest Branch. "Directory of Lumber Dealers Stocking British Columbia Woods." (pamphlet) 1917.

────────── "British Columbia Manufacturers of Forest Products Bulletin #19. (pamphlet) 1917.

Forest Service. "Annual Management Reports for the Prince George Forest District," 1927-1929, 1931, 1936, 1947, 1948, 1950-1953, 1956-1962.

────────── "Forest Industry Statistics of British Columbia," 1975.

Gazette, Volume XLVIII, Victoria, 1907.

Hansard. British Columbia Legislative Assembly, June 14, 1978.

Legislative Assembly, *Forest Act*, "Bill 14 as passed... June 29, 1978." Victoria: Queen's Printer, 1978.

Ministry of Forests. "Pulpwood Harvesting Area #5, Agreement of 6 June 1956 between the Minister of Lands, Forests and Water Resources and Cariboo Pulp and Paper Co. Ltd." (Document)

────────── *Annual Report*, 1974-1979 editions. Victoria: Queen's Printer.

────────── *Forest and Range Resource Analysis Technical Report*. Volumes 1 and 2. Victoria: Queen's Printer, March 1980.

Ministry of Recreation and Conservation. *Annual Report*, 1973, 1974, 1976/1977 editions. Victoria: Queen's Printer.

Royal Commission on the British Columbia Railway. "Proceedings at hearings..." Transcript, Provincial Archives, GR500, Box 12, Volume 42 and Exhibit 186.

Supreme Court of British Columbia: "Trial between Douglas Van Johnston, Gordon James Sales, Donald Drummond Stewart McKay and Dragon Investments Ltd., Petitioners, and West Fraser Timber Co. Ltd., Ketchum Investments Inc., TYSA Investments Inc., Janet Wright Ketchum and Christopher Paul Daniels." Transcript, Volumes 1-19, 1980-1981.

British Columbia Independent Loggers' Association. Brief presented to the Royal Commission on Forestry. 1975.

B.C. Logging News (magazine), January 1975.

B.C. Lumberman (magazine), September 1954.

British Columbia Natural Resources Conference. *Inventory of the Natural Resources of British Columbia*. 1964. Edited by D.G. Turner.

_____ *British Columbia Atlas of Resources.* 1965. Edited by J.D. Chapman and D.B. Turner.

Cameron, Colin. *Forestry... B.C.'s Devastated Industry.* Vancouver: Co-operative Commonwealth Federation, 1941.

Canadian Dimension (magazine), November 1971.

Canadian Forest Products. *A Salute to the Founders.* (booklet) Vancouver: undated.

Caves, Richard E., and Richard H. Holtan. "Outline of the Economic History of British Columbia: 1881-1951." In *Historical Essays on British Columbia,* edited by J. Friesen and H.K. Ralston. Toronto: McClelland and Stewart, 1976.

(The) Citizen (newspaper), Prince George, B.C. Dates are given in the text.

Coptithorne, Lawrence W. "Natural Resources and Regional Disparities: A Skeptical View." Discussion Paper 106, prepared for the Economic Council of Canada, February 1978.

Crockett, David. *Life of David Crockett.* New York: Perkins Book Co., 1903.

Fulton, Fred J. "The Final Report of the Royal Commission of Inquiry on Timber and Forestry, 1910." Bound with the *Sessional Papers* of the B.C. Legislative Assembly, 1910.

Galbraith, John Kenneth. *The New Industrial State.* Boston: Houghton Mifflin Company, 1967.

Glick, Wendell, editor. *The Recognition of Henry Thoreau.* Ann Arbor: University of Michigan Press, 1969.

Government of Canada. "Combines Investigation Act, Office Consolidation, 1978." Ottawa: Ministry of Supply and Services.

Grainger, M. Allerdale. "Forestry Progress in British Columbia." *The Timberman,* October 1920.

_____ *Woodsmen of the West.* 1908. Reprint. Toronto: McClelland and Stewart, 1964.

Haynes, Richard W. "Competition for National Forest Timber in the Northern, Pacific Southwest, and Pacific Northwest Regions." Portland, Oreg.: U.S. Department of Agriculture Northwest Forest and Range Experiment Station, January 1980.

The Herald (newspaper), South Fort George, B.C. All issues from #1, August 20, 1910, until the paper became the *Citizen* in 1917.

Hiballer Forest Magazine, Vancouver, B.C. All issues 1962-1979.

Holmens, Neil Bradford. "The Promotion of Early Growth in the Western Canadian City." B.A. Thesis, UBC, 1974

Howay, F.W. "Political History 1871-1913." In *Canada and its Provinces,* edited by Shortt and Doughty. Volume 21. Toronto: T.A. Constable, 1914.

Kavic, Lorne J., and Garry Brian Nixon. *The 1200 Days, a Shattered Dream.* Coquitlam, B.C.: Kaen Publishers, 1978.

Kenney, E.T. "Forestry in British Columbia." (pamphlet) Victoria: Queen's Printer, 1976.

Knowles, Joseph. *Alone in the Wilderness.* Boston: Small, Maynard & Company, 1913.

Lawrence, J.C. "Markets and Capital: A History of the Lumber Industry in British Columbia." M.A. Thesis, UBC, 1957.

Leopold, Aldo. *A Sand County Almanac.* Oxford University Press, 1966. Reprint. New York: Ballantine Books, 1970.

McBride, C.F. "Quality Control in the Forest Products Industry." Vancouver: Forest Products Laboratory, 1967.

McInnis, John. Interview on Pioneer Tape 19A, Prince George Public Library collection.

Machiavelli, Niccolo. *The Prince.* Translation copyright 1950. New York: The Modern Library.

Mead, Walter J. *Competition and Oligopsony in the Douglas Fir Timber Industry.* Berkeley and Los Angeles: University of California Press, 1966.

Morice, Father A.G. *History of the Northern Interior of British Columbia.* Toronto: Briggs, 1904. Reprint. Ye Galleon Press, 1971.

Mullholland, F.D. *The Forest Resources of British Columbia.* Victoria: King's Printer, 1937.

Mullins, Doreen. "Changes in Location and Structure in the Forest Industry of North Central British Columbia, 1909-1966." M.A. Thesis, UBC, 1967.

Murray, William H.H. *Adventures in the Wilderness.* Boston: Osgood & Company, 1869. Second printing, 1872.

Nagle, George Shorten. "Economics and Public Policy in the Forestry Sector of British Columbia." Ph.D. Thesis, Yale University, 1970.

Nash, Roderick, *Wilderness and the American Mind.* Revised Edition. New Haven: Yale University Press, 1972.

New Democratic Party of British Columbia. *Policies and Peoples.* Vancouver: 1976 edition.

Northwood Pulp and Timber Ltd. Brief submitted to the Royal Commission on Forest Resources. August 14, 1975.

————— Brief submitted to the British Columbia Hydro and Power Authority. May 15, 1976.

————— Northwood News I:1. December 1980.

Orchard, C.D. "Forest Administration in British Columbia, a Brief for Presentation to the Royal Commission on Forestry." Victoria, 1945. Mimeographed.

————— "Forest Management." (pamphlet) Third revised edition. Victoria: Queen's Printer, 1954.

Ormsby, Margaret A. *British Columbia: A History.* Toronto: Macmillan Co. of Canada, 1958.

Patullo Papers (letter from the Fraser River Syndicate), December 14, 1920. Provincial Archives, Victoria, B.C.

Pearse, Peter H. *Timber Rights and Forest Policy in British Columbia, the Report of the Royal Commissioner on Forest Resources.* Victoria: Queen's Printer, 1976.

————— Proceedings before the Royal Commission on Forestry. Transcript, volumes 1-7, 1975.

————— *Timber Appraisal, the Second Report on the Task Force on Crown Timber Disposal.* Victoria: Queen's Printer, July 1974.

————— "Does Competition Matter?" Speech, January 9, 1980.

————— "A Review of the Ministry of Forests White Paper on Crown Timber Pricing." Speech, August 1980.

Persky, Stan. *Son of Socred.* Vancouver: New Star Books, 1979.

Pope, Dudley. *The Great Gamble: Nelson at Copenhagen.* London: Weidenfeld and Nicholson, 1972.

Potter, Dale R., with Kathryn M. Sharpe and John C. Hendee. *Human Behavior Aspects of Fish and Wildlife Conservation, an annotated bibliography.* Portland Oreg.: USDA Northwest Forest and Range Experiment Station, 1973.

Prince George Leader (newspaper), September 16, 1921.

Reed, F.L.C., & Associates. *Forest Management in Canada.* Volumes 1 and 2. Ottawa: Forest Management Institute, January 1978.

————— *Forest Management Expenditures in Canada Compared to Taxes* Vancouver: F.L.C. Reed & Associates, 1980.

Robbin, Martin. *Pillars of Profit.* Toronto: McClelland and Stewart Ltd., 1973.

Ross, William R. "British Columbia Forest Policy." Speech to the British Columbia Legislative Assembly, February 1912. Reprint. Pamphlet in the B.C. Legislative Assembly Library.

Runnalls, Reverend R.E. *A History of Prince George.* Vancouver: Wrigley Printing Co. Ltd., 1946.

Sherman, Paddy. *Bennett.* Toronto: McClelland and Stewart, 1966.

Silk, Leonard, editor. *Capitalism: The Moving Target.* New York: New York Times Book Co., 1973.

————— *Economics in Plain English.* New York: Simon & Shuster, 1978

Sloan, Gordon McF. *Report of the Commissioner... Relating to the Forest Resources of British Columbia.* Victoria: King's Printer, 1945.

_____ *Report of the Commissioner Relating to the Forest Resources of British Columbia.* Volumes 1 and 2. Victoria: Queen's Printer, 1965.

Sommers, Robert E. Speech to the Legislative Assembly, February 1956, in booklet form in the B.C. Legislative Assembly Library.

Spokesman-Review (newspaper), Spokane, Washington, July 29, 1948.

Stevens, G.R. *Canadian National Railways.* Volume 2. Toronto: Clarke, Irwin & Company Ltd., 1962.

Taylor, Geoffrey. *Timber: History of the Forest Industry in B.C.* Vancouver: J.J. Douglas Ltd., 1975.

Thoreau, Henry David. *Walden.* New York: Signet edition, 1961.

Truck Logger (magazine), August 1954.

Vancouver Sun (newspaper). Dates are given in the text.

Walker, Russell R. *Politicians of a Pioneering Province.* Vancouver: Mitchell Press, 1969.

Western Lumberman (magazine), June 1907, April 1910.

White, Stewart Edward. *The Blazed Trail.* Originally serialized in *McClure's Magazine,* 1901. Garden City, N.Y.: Doubleday, Page and Company, 1902.

_____ *Camp and Trail.* Garden City, N.Y.: Doubleday, Page and Company, 1913.

_____ *The Riverman.* Originally serialized in *McClure's Magazine,* 1907-1908. Garden City, N.Y.: Doubleday, Page and Company, 1908.

_____ *The Rules of the Game.* Garden City, N.Y.: Doubleday, Page and Company, 1910.

Williston, Ray. Transcript of interview, Oral History Section, B.C. Provincial Archives, #1375.

_____ "The Pulp and Paper Industry of the Northwest; Past, Present and Future." Speech at the B.C. Institute of Technology, June 22, 1979.

Wood, Nancy. *Clearcut.* San Franscisco: Sierra Club Books, 1971.

Index

Abitibi Paper, 127
Adventures in the Wilderness, 135
Alexander Lumber Company, 120-21
Alexandra Forest Products, 100
Aleza Lake, 33, 35, 123-24
Aleza Lake Mills, 31
Anderson, Ivan, 119-25
Angier, Bradford, 135
Apollo Forest Products, 77, 119, 122, 176
Apsey, Mike, 181
Asling, John, 183
B. & B. Logging, 121
Babine Forest Products, 122, 150, 176
Backman, Bill, 146
Balco Industries, 126
Balfour-Guthrie, 123, 129-31, 145
Barrett, Dave, 142, 147, 148, 150
Batcheller, Willis T., 64
B.C. Logging News, 149
B.C. Lumberman, 145, 148
Bear Lake, 131
Bell, Ian, 111, 128
Bend Lumber Company, 34
Bennett, Joe, 170
Bennett, W.A.C., 64, 71, 79, 85, 97, 141
Bentley, Poldi, 96, 97
Bernhardt brothers' mill, 128
Big Bend Lumber Company, 79
Blackburn and Hasafield's mill, 124
Bogue and Brown's mill, 13, 22, 27
Boise Cascade Corporation, 170
Bonner, Robert, 79
Bowaters-Brathurst, 99, 118
Bowater's Canadian Corporation, 118
Bowell, Gary, 148
Bowie, Alex, 69, 72
Bowron Lake Lumber Company, 126
British Columbia Forest Products, 80, 100, 108, 111, 112, 113, 142, 156
British Columbia government
   Fish & Wildlife Branch, 136, 137, 138, 139
   Forest Act. *See* Forest Act (B.C.)
   Forest Policy, 11, 36-39, 40-43, 54-57, 82-89, 96-99, 111, 122, 124, 137, 139, 140-53, 154-63, 172-78. *See also* Close utilization policy; Competition for timber; Quota; Sustained yield; Tenure; Royal Commissions on Forestry
   Forest Service, 17-19, 36-43, 77, 78, 83, 86, 89, 93, 102, 117, 122, 135-39, 142, 151, 154, 158, 172, 180, 183. *See also* Prince George Forest District
   Parks, 136, 138-39
   Timber Products Stabilization Act 147-48
British Columbia Railway, 145, 146, 148, 150. *For early history see* Pacific and Great Eastern Railway
British Columbia Research Council, 148
British Columbia Resources Investment Corporation, 127
Brookmere Logging, 117
Brown, Vic, 99
Brownmiller Lumber Company, 126
Brownridge, Gordon, 111
Brynlesen, Bern, 115
Buck River Lumber Company, 118
Bulkley Valley Forest Industries, 111-12, 116, 118, 122, 150
Burns Lake, 114

Caine, Martin, 120-21
Cameron, Colin, 51, 54-57
Camp and Trail, 135
Canadian Car, 112-13
Canadian Cellulose Ltd., 144, 149
Canadian Dimension, 140
Canadian Forest Products, 96-99, 114, 119, 157, 158
Canadian Hydrocarbons Ltd., 126
Canadian National Railway, 37, 40, 55, 58, 62-63, 72, 145. *For early history see* Grand Trunk Pacific
Canadian Northern Railway, 28
Canadian Pacific Railway, 10
Canadian Paperworkers Union, 150
Capitalism: The Moving Target, 141
Cariboo Lumber Company, 33, 34
Cariboo Pulp and Paper Ltd., 99, 107, 142
Carney, Pat, 83
Carrier Lumber, 59, 77, 175
Cattermole, Robert, 100
Celanese Corporation, 149
Central Fort George *Tribune*, 13, 14
Chabot and Monteith mill, 33
Chamberlin, Edwin H., 88
Chambers, Alan, 132
Chetwynd, 124, 126
Chip direction policy, 98
Chip-N-Saw, 112-13
Church, Percy, 110-11, 116, 118
Clarke, Nick, 12, 21
Clear Lake Sawmills, 110-11, 131, 147
Clinton, Russ, 127
Close utilization policy, 83-84, 97, 108, 111. *See also* Tenure, "Third band"; Quota
Competition and Oligopsony in the Douglas Fir Lumber Industry, 89-90
Competition for timber, 76-78, 87-91, 99-100, 104-105, 124, 141, 151, 160-62, 173
Conflicts between forest users, 132-39
Consolidated Brathurst, 118. *See also* Bowaters-Brathurst
Cooke, William, 13, 21
Cook Lumber Company, 34
Co-operative Commonwealth Federation, 54-57
Cooper-Widman, 123
Coptithorne, Lawrence, 182
Cornell Sawmills, 79
Council of Forest Industries, 63, 117, 129, 136, 142, 143, 145, 148, 161, 168
Cranbrook Sawmills, 33
Crestbrook Pulp and Paper, 99-100, 107
Crockett, Davy, 133
Cromwell, Fred T., 12
Crown Zellerbach, 136, 143, 170
Crows Nest Industries, 99-100, 126
Currie, Don, 131

Daishowa, 111, 127
Dalskog, Ernie, 46
Davenport, David, 115
Davis, H.V. and J.V., mill, 58
DeGrace, Larry, 77, 85
Dewey, 79
Dezell, Garvin, 74
Diefenbaker, John, 83
Dilworth, Tom, 123-24
Dome Creek, 28
Domtar mill, 126
Draesaeke, Gordon, 143

Eacom, 131
Eagle Lake Sawmills, 31, 35, 52, 59, 108, 109, 116, 118, 120, 148
Economic Council of Canada, 182
Ellett, Jack, 148
Ellison, Rollie, 128
Emerson, Ralph Waldo, 133
Endako, 76, 112
Ernst Forest Products, 113, 147
Ernst, John, 84, 113, 147
Etter & McDougall Lumber Company, 33
Eurocan Pulp and Paper, 127, 150, 175, 176
Eversfield, Charles, 79
Farrow, Moira, 65
Farstad, Al, 99
Feldemuhle, 99
Ferguson Lake Sawmills, 145
Fernow, Bernard, 40
Fichtner's mill, 116, 118
Fieber, Fred, 69, 70
Findlay Forest Industries, 100, 111, 126
Fink Sawmills, 176
Floyd, Doug, 127
Flumerfelt, A.C., 20, 21
Flynn, Don, 111
Flynn, Jerry, 111
Foley, Welch, and Stewart, 62
Foreman Lumber Company, 33, 34
Forest Act (B.C.), 11, 22, 77, 89, 110-11, 172-78
Forest Associations (regional), 77-78
Forest Policy. *See* British Columbia government, Forest Policy
Forest Products Laboratory, 108
Forestry...B.C.'s Devastated Industry, 51, 54-57
Fort Fraser, 10
Fort George, 10, 12-13, 14, 23, 24
   Central Fort George, 12, 14, 21
   South Fort George, 12-13, 14, 21
Fort George Trading and Lumber Company, 21
Fort McLeod, 10
Fort St. James, 10, 63, 111, 122, 150, 164
Fort St. John, 64, 107
Fraser City, 14
Fraser River Syndicate, 29-31
Fraser, Simon, 10
Fulton, Fred J., 11, 16-19, 40, 91
Gale and Trick's mill, 3
Gairns, Harry, 145, 146
Gallagher, Bob, 66, 94
Gardner Building Supplies, 126
Geddes, Gordon, 111
Georgia-Pacific Corporation, 170
Gibson, Gordon, 79
Ginter, Ben, 100
Giscome, 15, 24, 26, 27, 31, 45, 47, 51, 116, 149
Goodład, Roger, 137, 145
Grainger, Bill, 77
Grainger, M. Allerdale, 17
Grand Forks, 107
Grand Trunk Pacific, 11-13, 20-23, 33. *See also* Canadian National Railway
Gray, H. Wilson, 79
Gregory Industries, 176
Greater Peace Forest Products, 107
Green Lake Lumber, 126
Greenwood, 107
Guilford Lumber, 45

Habriele, Eugene, 72
Haig-Brown, Roderick, 136
Hamilton, A.C., 12
Hammond, George J., 12
Hanaford's mill, 128
Hansard Lake Lumber, 31
Hanzlick formula, 42
Hawker Siddley, 112-13
Hawthorne, Nathaniel, 133
Hazelton, 76, 148
Helco Forest Industries, 121, 176
History of the Northern Interior of British Columbia, 10
Holst, Jake, 70, 94
Honshu Pulp and Paper, 99
Houston, 102, 118, 150, 164-66, 175
How to Live in the Woods on Pennies a Day, 135
Hutton, 26
Independent Chip Producers, 147
Industrial Forestry Service, 77, 85, 145
Integrated Wood Products, 126
Intercontinental Pulp, 107
Interior Spruce Mills, 128
International Woodworkers of America (IWA), 52, 94-95, 142, 147, 150, 154, 180, 182. *See also* Local 1-424
International Workers of the World (IWW), 52
Inventory of the Natural Resources of British Columbia, 38, 43
Jacobson Brothers Sawmill, 145
Johnson, Bill, 170
Johnson, Byron, 64, 66
Johnston, Doug, 126
Jordan, Irene, 21
Ketchum, Bill, 125-26
Ketchum, Pete, 125-26
Ketchum, Sam, 84, 125-26, 127
Killy, George, 119, 121
Killy, Ivor, 119, 145
King, Bill, 147
Kitimat, 102
Kluskus Indian bands, 149, 150
Knowles, Joe, 134
Koerner, Leon, 42
Kootenay Forest Products, 144, 146
L. & M. Lumber, 119
Lakeland Mills, 119, 121, 123-25, 149, 161, 176
Lamb Brothers, 121
Leontif, Wassily, 141
Leopold, Aldo, 134-35
Liersch, John, 96, 97, 98
Lignum Ltd., 126, 176
Lloyd Brother's mill, 110-11
Lloyd, Howard, 111, 146
Local 1-424, 46, 66-74, 94, 95, 106, 108-9
Longworth, 105
Lord, Arthur A., 73
Lumber production. *See* Prince George Forest District, Statistics on production, mills, and harvest
Lutton, John, 120
Lyle, Lorne, 124
Lyle Lumber Company, 33, 124
McBride, 27, 59

McBride, Richard, 10-12, 16, 22, 62, 64
McBride Timber, 128
McDonald, Alex, 71
McGregor, 118
McInnis, John, 13, 72
Mackenzie, 85, 100-102, 108-9, 111, 142, 164-66
McLean Lumber Company, 33
McLeod, W.A., 126
MacMillan Bloedel, 96, 98-99, 117, 126, 131, 144
McTaggert-Cowan, Ian, 136
Mahood, Ian, 82, 147
Marketing of lumber, 168-70
Marshall, Harry, 138
Martin, John, 111
Martin, Pat, 111
Martin's mill, 138
Mead Paper Company, 112, 114-19. *See also* Northwood Pulp and Timber
Mead, Walter J., 89-90
Mennonites, 27
Midway Terminals, 104
Mitsubishi, 99
Moffat, A.B., 72
Moffat, Harold, 84-85
Mogensen, Tage, 72
Morice, Father A.G., 10, 40
Morrison, Harry, 124
Mueller, Carl, 70
Muirhead Machinery, 112
Mulholland, F.D., 40-41, 55, 57
Munro, Bill, 72
Munroe, Craig, 108
Murray, William H.H., 135
Nagle, George, 111
Nance's mill, 118
Nash, Roderick, 133
National Forest Products, 104, 106, 115-16
Natural Resources Security Company, 12, 13, 20
Nazko Indian Band, 149, 150
Neal, Harry, 44, 70
Nechako Lumber, 119
Netherlands Overseas Mills, 77, 110, 111, 121, 128-31
New Demoncratic Party (NDP), 54, 65, 138, 139, 140-53, 176
Newlands, 118
Newlands Sawmills, 33
Nixon, Richard, 142, 147
Noranda Mines Ltd., 106, 109, 112, 114-19
North Central Plywood, 77, 111, 119, 146
Northern Development Company, 12, 13
Northern Forest Products, 94
Northern Interior Lumbermen's Association (NILA), 36, 57, 66-74, 94-95, 108-9
Northern Lumber and Mercantile Company, 21, 24, 26, 27
Northern Planers, 73, 120
Northern Spruce Mills, 128
Northland Spruce Lumber Company, 34
Northwood Pulp and Timber (including Northwood Mills), 99, 106, 108-9, 111-12, 114-19, 126, 131, 148, 149, 150, 176, 183
Ocean Falls Corporation, 143
Oliver, John, 31, 62
Oliver Sawmills, 120
Orchard, C.D., 39, 45, 55, 80
Oregon Pacific, 123

Pacific and Great Eastern Railway, 20, 22, 23, 24, 29, 31, 33, 40, 61, 62-65
Pacific Inland Resources, 126, 176
Pacific Northern Railway, 64
Patchett, A.C., and Sons, 126
Patullo, T.D., 29
Peace River Block, 10, 63, 66
Peace River Kraft Mills, 107
Pearse, Peter, 16, 43, 55, 81, 90-91, 97, 146, 148, 151, 154-63, 174-75, 176, 183
Pearson, Lester, 142
Peden, William, 13, 14, 21
Penny, 28, 33, 45, 49
Penny Lumber, 31
Penny Spruce Mills, 26, 72, 79, 118
Perry, Harry, 64
Plant, Harry, 64
Plant, Ralph S., 123
Plateau Mills, 144-45, 151, 175
Polar Forest Industries, 129, 131
Powell River Company, 16
Powis, Alfred, 115
Prentice, John, 96
Price Brothers, 99, 107, 111
Prince George, 20, 23, 24, 27, 31, 33, 34, 36, 47, 51, 58-59, 62, 63, 64, 65, 66-74, 76, 84-85, 91, 96-103, 107-9, 112, 114-25, 128-31, 147, 164-66. *For early history see* Fort George
Prince George *Citizen,* 21, 24, 27, 49, 64, 67, 69, 72, 73-74, 78, 97, 143, 147, 183
Prince George Forest District
    Administration, 24, 33, 36-39, 40-41, 45, 49, 58-61, 78-79, 88-89, 92-94, 104-7 111
    Statistics on lumber production, mills, and harvest, 24, 27, 28, 31, 33, 44, 45, 58, 59, 63, 78, 82, 86, 92-93, 104-9, 113, 143
Prince George *Leader,* 31
Prince George Planing Mills, 67, 72
Prince George Pulp, 98-99, 107, 108, 146
Prince Rupert, 12, 40, 76
Pulp and Paper Workers of Canada, 146, 150
Pulp mills, 28-30, 63, 96-103, 104-9, 111, 114-19, 123, 148, 149, 150. *See also names of individual companies*
Q.M. Industries, 112
Quesnel, 27, 31, 33, 40, 51, 58, 59, 62, 66, 71, 99, 113, 125-28, 142, 145
Quesnel Supply Company, 126
Quota, 60, 77, 86-91, 92, 103, 104-5, 124, 141, 172-73, 176-77

Racism, 27, 31-33
Raush Valley, 27
Red Mountain Lumber Company, 28, 33
Red Rock, 118
Reed, F.L.C. (Les), 181
Reed, F.L.C., & Associates, 38
Reed Group, 98-99
Remo, 28
Robbins, Ralph, 104
Robinson, Joan, 88
Richardson, George, 120
Rigler, Buster, 121
Rim Forest Products, 148-49
Rockefeller, David, 141-42
Ross, William R., 18-19, 177-78

191

Rossi, Herman J., 14
Royal Bank, 110-11, 149
Royal Commissions on Forestry
   1910 Royal Commission on Forestry
      (Fulton Report),11, 16-19, 40, 91
   1945 Royal Commission on Forestry
      (Sloan 45), 41, 54-57
   1956 Royal Commission on Forestry
      (Sloan 56), 41, 76-77, 80-81, 88
   1976 Royal Commission on Timber Rights
      and Forest Policy (Pearse Report), 43,
      90-91, 154-63, 174-75, 176, 183
RoyNat, 110-11
Rustad Brothers' mill, 72, 77, 121, 169, 176
Rustad, Jim, 169

Safford, Edwin R., 120
Sand County Almanac, 134
Scoffield, Van, 145, 168
Seaboard Lumber Sales, 66, 131, 169
Sekora, Mike, 70
Shaw, Dean, 111
Shelley, 23, 33
Shelley Sawmills, 31, 116
Sierra Club, 138
Sierra Pacific, 126
Silk, Leonard, 140
Silvican Resources, 77
Sinclair Mills (town), 45, 46, 51
Sinclair Spruce Mills, 51, 104, 115, 117, 118,
   120-21
Sinclar Enterprises, 119-25
Six Mile Lake Sawmills, 46
Skagit Industries, 169-70
Sloan, Gordon McFarlane, 41, 54-57, 76-77,
   79, 80-81, 88, 163
Smith, Norman M., Lumber Company, 128
Smithers, 114, 139, 176
Sommers, Robert, 79, 80, 85
Sopow, Eli, 183
Special Licences, 11
Spurr, Roy, 26, 34, 52, 108, 120
Squamish, 107
Sterling, Bill, 149
Stevenson, Robert Louis, 133
Stewart, Bob, 119-25
Stokes, John, 96
Strom, Lars, 120
Summerland Box, 115
Sustained yield, 37, 39, 54-57, 60-61, 82, 86,
   180
Swankey, Gordon, 111
Swetman, William J., Ltd.(and associated
   companies), 126

Tahsis, 176
Takla Forest Products, 77, 98
Tenure
   Forest Licences, 161, 176
   Hand Logger's Licences, 11
   Public Sustained Yield Units, 57, 59-61,
      77-78, 86-87, 89, 96, 97, 104-5, 176
   Pulpwood Harvesting Agreements, 96-100,
      176
   Special Licences, 11
   Special Sale Areas, 88
   "Temporary" tenures, 146
   "Third band," 83-84, 103, 111
   Timber Licences, 90, 161
   Timber Sale Harvesting Licences, 111
   Timber Supply Areas, 57, 176

Tree Farm Licences, 57, 60, 76, 79, 80,
   154-76
Transfer of tenures, 110-11
Terrace, 33
Tete Jaune Cache, 22
"Third band." *See* Tenure.
Thrasher, F.G., Lumber Company, 34
Thursday Lumber, 106
Tilstra, Fred, 73-74
Timber Supply Areas, 57, 176
Trapping Lake Sawmill, 59
Triangle Pacific, 147
Trick, Ambrose, 124
Trick, S.B., Lumber Company, 123-24
Truck Logger magazine, 76
Truck Logger's Association, 136, 145, 147
Tulameen Forest Products, 117
Two Mile Planing Company, 125-26
Tyhurst, Robert, 30

Unionization, 45-46, 66-74
United Grain Growers, 26
Upper Fraser, 45
Upper Fraser Lumber Company, 27, 31, 104,
   109, 115, 118, 120, 121
United Pulp, 107
United States Forest Service, 11, 90, 163

Valemount, 106
Vancouver Sun, 65, 136
Vanderhoof, 23, 27, 33, 66, 76, 133
Van Drimmelen, Nick, 110-11, 128-30
Vick Brothers' mill, 45

Waterland, Tom, 89, 155, 156, 163, 172-78,
   181, 182
Webb, Howard, 69, 72
Webster, Arnold, 72-72
Weldwood, 60, 77, 99, 107, 108, 111, 147,
   148, 175, 176. *For early history see*
   Western Plywood
Wenner-Gren, Axel, 64, 85, 100
West Fraser Mills, 77, 125-28, 170, 176, 181
   Western Lumberman, 11-12, 14
Western Plywood, 59, 76. *See also*
   Weldwood
Weyerhauser (Canada) Ltd., 116
Wheeler, Gray, 149
White, Stewart Edward, 135
Whitmer, John, 128-31
Wilderness and the American Mind, 133
Williams, Bob, 87, 89, 138, 139, 140-53, 173,
   175, 183
Williams, Viv, 136
Williams Lake, 33, 62, 71, 126
Williston, Ray, 58, 64, 72, 74, 79, 80, 81,
   82-88, 89, 90, 94, 95, 96-103, 108,
   110-11, 117, 124, 129, 154, 156, 173, 182
Winther, Chris, 111
Winther, Eric, 111
Wood, Robert, 89, 161, 174-75
Wright, Tom, 84, 96
Wright Forest Products, White Trucking,
   and associated companies, 126

Yip, Roy, 69
Young, Bill, 43, 78, 93, 137, 145, 180-81
Young, Ted, 146

Zimmerman, Adam, 106, 115-19